# Usborne
# 100 Science Experiments

Georgina Andrews and Kate Knighton

Designed by Zoe Wray and Tom Lalonde

Models by Katie Lovell
Illustrations by Stella Baggott
Photography by Howard Allman
Edited by Jane Chisholm

Consultants: Dr. Catherine Cooper and Liz Lander,
Science Curriculum Coordinator, Roehampton University

# Contents

# Internet links

Throughout this book we have recommended fun websites where you can do experiments online and find more experiments to do at home. For links to all the recommended websites in this book, go to the Usborne Quicklinks Website at **www.usborne-quicklinks.com** and enter the keyword "experiments".

## How to use Usborne Quicklinks

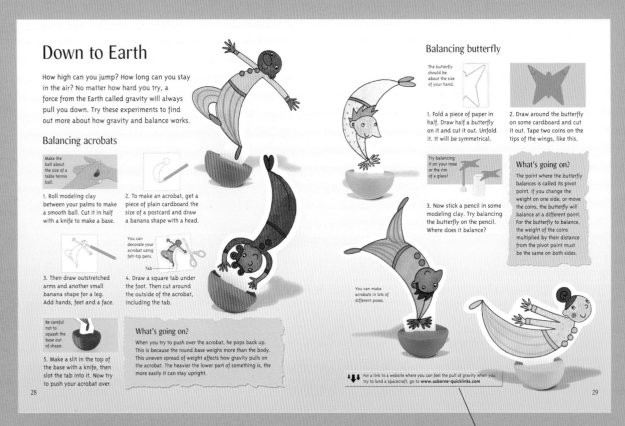

1. Look for the "Internet links" boxes on the pages of this book. They contain descriptions of the websites you can visit.

 For a link to a website where you can feel the pull of gravity when you try to land a spacecraft, go to **www.usborne-quicklinks.com**

2. In your computer's web browser, type the address **www.usborne-quicklinks.com** to go to the Usborne Quicklinks Website.

3. At the Usborne Quicklinks Website, type the keyword for this book: "experiments".

4. Type the page number of the link you want to visit. When the link appears, click on it to go to the recommended site.

The links in Usborne Quicklinks are regularly updated, but occasionally you may get a message that a site is unavailable. This might be temporary, so try again later, or even the next day. Websites do occasionally close down and when this happens, we will replace them with new links in Usborne Quicklinks. Sometimes we add extra links too, if we think they are useful. So when you visit Usborne Quicklinks, the links may be slightly different from those described in the book.

# Websites to visit

Here are some examples of the many things you can do on the websites recommended in this book:

- Play with an online kaleidoscope
- Print out a froggy flick book
- Test friction with some wind-up toys
- Have a go at some building challenges
- Visit an alien juice bar to find out about acids and alkalis
- Experiment with an online paper plane simulator
- Design a roller coaster

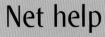

# Staying safe online

Make sure you follow these simple rules to keep you safe online:

- Children should ask an adult's permission before connecting to the Internet.
- Never give out personal information about yourself, such as your real name, address, phone number or school.
- If a site asks you to log in or register by typing your name and address, children should ask permission from an adult first.
- If you receive an email from someone you don't know don't reply to it. Tell an adult.

**Adults** - the websites described in this book are regularly reviewed and updated, but websites can change and Usborne Publishing is not responsible for any site other than its own. We recommend that children are supervised while on the Internet, that they do not use Internet chat rooms and that filtering software is used to block unsuitable material. You can find more information on Internet safety at the Usborne Quicklinks Website.

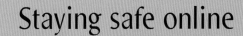

# Net help

For information and help using the Internet, go to the Net Help area on the Usborne Quicklinks Website. You'll find information about "plug-ins" — small free programs that your web browser needs to play videos, animations and sounds. You probably already have these, but if not, you can download them for free from Quicklinks Net Help. You can also find information about computer viruses and advice on anti-virus software to protect your computer.

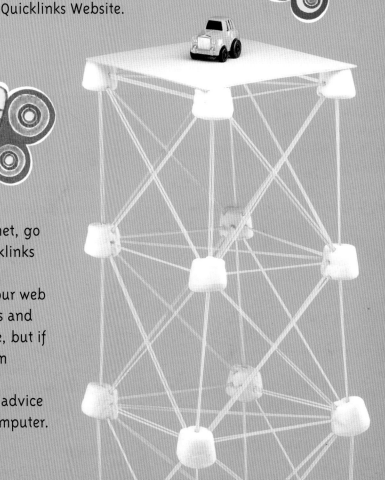

# What you'll need

You can get started on the experiments in this book right away. Most of them use cheap and simple materials you'll probably already have at home. Here are lots of things you'll find useful. It's a good idea to read through each experiment before you start, so you can collect together everything you need.

## Simple stationery

Many of the experiments use simple stationery and art materials such as paper, pens, pencils, ball-point pens, paints, paperclips, thumbtacks, glue, clear tape, a hole puncher, rubber bands, string, poster tack, modeling clay, balloons and cardboard.

## Kitchen things

Kitchen utensils and equipment, such as bowls and silverware, are needed in some experiments. Items such as foil, food wrap, wax paper, paper towels, toothpicks and dishwashing liquid will also be useful. For some experiments you will need some basic food ingredients.

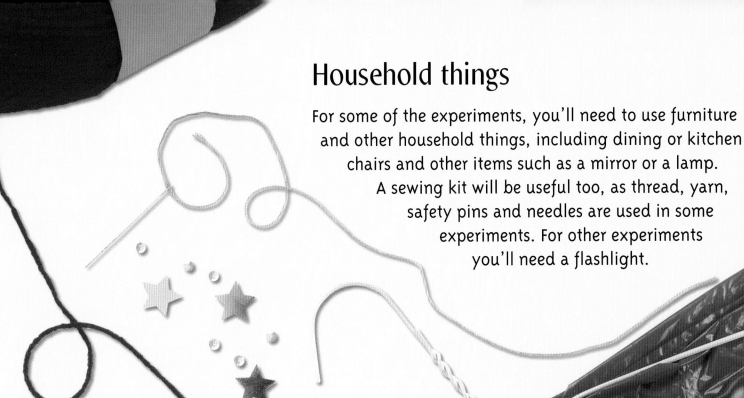

# Household things

For some of the experiments, you'll need to use furniture and other household things, including dining or kitchen chairs and other items such as a mirror or a lamp. A sewing kit will be useful too, as thread, yarn, safety pins and needles are used in some experiments. For other experiments you'll need a flashlight.

# Recycling

Save old food containers — but wash them out first. Jars, bottles (glass and plastic), boxes and plastic tubs will all come in handy. You'll also be able to make use of old newspapers, shoeboxes, old pantyhose and plastic bags. All kinds of packaging can be a great source of free cardboard and plastic. You might also want to save candy wrappers and wrapping paper, which you can use to decorate some of the things you make.

# Special equipment

There are only a few things that might be harder to find. When these are needed, there are suggestions for where to find them. They are all available inexpensively.

# The power of the Sun

The Sun is the main source of heat and light on Earth. Here are some experiments to try on a sunny, summer's day. You can use the Sun's heat to cook things, and see how the shadows cast by the Sun change with the time of day.

## Cooking in the Sun

1. Line a large bowl with kitchen foil. Then press a piece of poster tack down in the middle of the bowl.

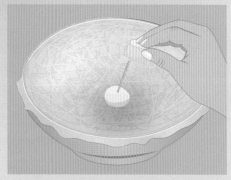

2. Put a marshmallow on the end of a toothpick. Push the other end of the toothpick into the poster tack.

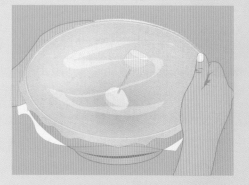

3. Cover the top of the bowl with clear food wrap. Then put the bowl outside in a sunny place.

4. Use stones to prop up the bowl. Position it so that the inside is facing the Sun. Leave it for about 15 minutes.

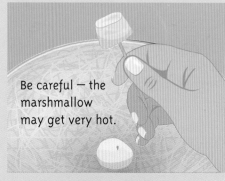

Be careful — the marshmallow may get very hot.

5. The marshmallow should start to melt. If it hasn't, leave it for another 15 minutes and check again.

## What's going on?

The clear food wrap lets sunlight into the bowl, while also trapping heat from the Sun. The foil reflects the light and heat around the bowl and onto the marshmallow. This heats it up. Because the air in the bowl is trapped, it gets hotter and hotter, which also helps to cook the marshmallow.

For a link to a website where you can watch a short movie about the Sun, go to **www.usborne-quicklinks.com**

# Make a Sun circle

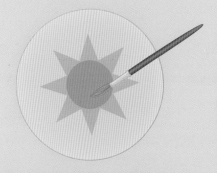

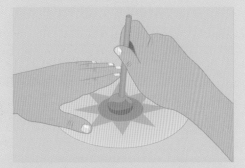

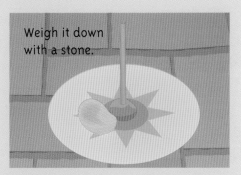
Weigh it down with a stone.

1. Draw around a plate on some cardboard to make a circle. Cut it out. Then paint it and draw a Sun shape on it.

2. Press a lump of poster tack into the middle of the circle. Cut a straw in half and push it into the poster tack.

3. Now put the circle in a sunny place outside, where it won't be overshadowed by buildings or trees.

Don't move the circle. Keep it in exactly the same position.

The shadow will change position over the course of the day.

4. Use a ruler to draw a line where the straw's shadow falls. Check the time and mark it on the line.

5. Mark the new position of the shadow each hour. What do you notice about the length of the lines?

## What's going on?

As the Earth spins during the day, the Sun appears to move in the sky. The shadows are longer before and after noon, because the Sun is lower in the sky. At noon the Sun is at its highest point, so it casts the shortest shadow.

The Sun circle works a bit like a sundial, which people used for telling the time before clocks were invented, but it's not as accurate. To show the time accurately, the straw part has to be set at an angle, depending on the time of year and what part of the world you live in.

Don't look directly at the Sun; it can damage your eyes.

You could add glitter and stars for decoration.

You could try using the Sun circle to track the Sun at a different time of year. Does it change its pattern?

3 pm
2 pm
1 pm
12 noon
11 am
10 am
9 am

# Lighting effects

Light travels in straight lines called rays. But rays can change direction if they're reflected by things such as mirrors or water. Here you can discover how to change the way light is reflected — with some surprising results.

## Make a kaleidoscope

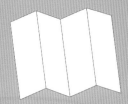

1. Fold a postcard in half, so that the shorter edges meet. Then fold it in half the same way again. Open it out.

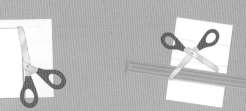

Clear plastic can be found in some packaging.

2. Find some clear, stiff plastic. Cut a piece the same size as the postcard and lay it on top of the postcard.

3. Score lines on the plastic, on top of the postcard folds, using scissors and a ruler. Put the plastic aside.

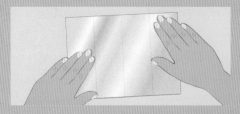

4. Now cut out a piece of foil the same size as the postcard. Glue it onto the postcard and smooth it out with your fingers.

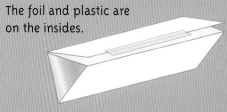

The foil and plastic are on the insides.

5. Next, lay the plastic on the foil and fold the postcard into a triangular tube. Tape the fourth flap over the first.

6. Cut out a piece of tracing paper larger than the end of the tube. Draw patterns on it using felt-tip pens.

7. Look through one end of the tube and hold the paper to the other end. Point it to the light and move the paper around.

## What's going on?

Light shines through the decorated tracing paper into the tube. The plastic-covered foil sides act like mirrors, reflecting the light. Each side also reflects the light that reflects from the other sides. All these different reflections create interesting patterns of colored light.

 For a link to a website where you can make fun patterns with an online kaleidoscope, go to **www.usborne-quicklinks.com**

# Fountain of light

Use a pen to widen the hole.

This effect is easier to see in a dark room.

1. Find a flashlight and a large plastic bottle. Pierce the bottle halfway up with a thumbtack. Hold your finger over the hole and fill the bottle with water.

2. Position the hole, facing a sink. Shine a flashlight through the bottle from behind the hole. Take your finger away. The stream of water will light up.

## What's going on?

You might expect the flashlight beam to pass through the bottle and out the other side. Instead, the light gets trapped inside the stream of water. It reflects off the sides of the stream, and bends with it into the sink.

# Make a pinhole projector

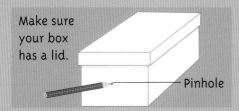

Make sure your box has a lid.

Pinhole

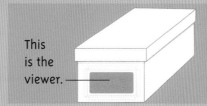

This is the viewer.

1. Push a thumbtack into the middle of one end of a shoe box to make a hole. Push a pencil in to widen it.

2. Cut out a rectangular window at the other end of the box. Tape wax paper over it.

3. Cut another piece of wax paper big enough to cover the lightbulb end of a flashlight.

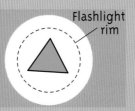

Make sure the triangle is slightly smaller than the end of the flashlight.

Flashlight rim

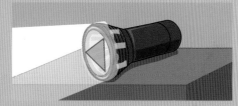

The image will be blurry.

4. Draw a triangle on the paper. Use felt tip pens to fill the triangle in dark green or blue and outline it in black.

5. Tape the paper to the end of the flashlight. Switch on the flashlight and place it on a surface in a dark room.

6. Stand about 3 feet from the flashlight. Look through the viewer, pointing the pinhole at the light. What do you see?

## What's going on?

The light from the flashlight passes through the pinhole and onto your viewer. Light rays from the top of the flashlight hit the bottom part of the viewer, and rays from the bottom hit the top of the viewer. These rays cross over when they pass through the pinhole, so you see the image of the triangle upside down.

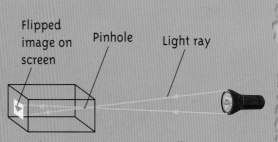

Flipped image on screen    Pinhole    Light ray

# Shadow show

Shadows are just areas where light has been blocked. You can put on a shadow puppet show by using puppets to block the light. Their shadows fall onto a screen and can be seen by the audience watching from the other side of the screen. Try it here.

Monkey puppet

1. To make the screen, take a big sheet of white paper and paint tree trunks in black up the sides of it.

2. Add curved palm leaves at the tops of the trunks. Paint flowers and grass along the bottom.

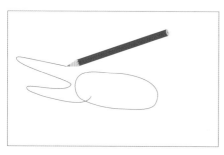

3. To make a crocodile puppet, draw a long body shape on a piece of cardboard. Then add jaws and a tail.

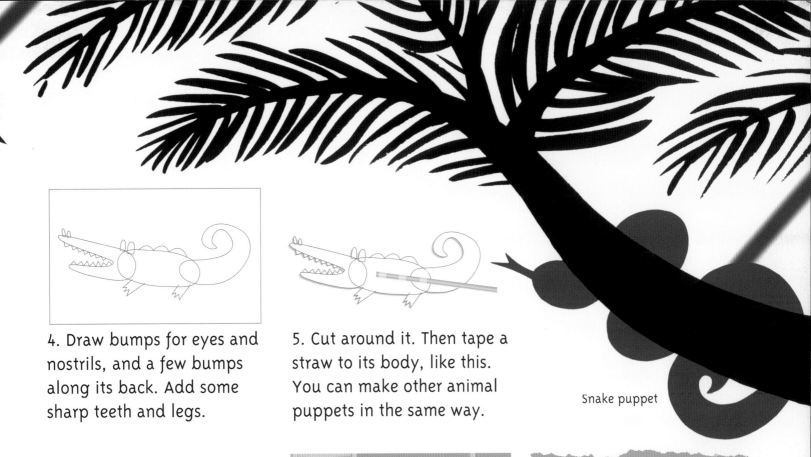

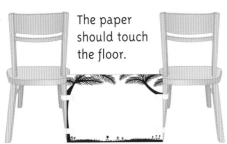

4. Draw bumps for eyes and nostrils, and a few bumps along its back. Add some sharp teeth and legs.

5. Cut around it. Then tape a straw to its body, like this. You can make other animal puppets in the same way.

Snake puppet

The paper should touch the floor.

6. Use masking tape to attach the screen between two chairs, with the painted side facing the front.

7. Darken the room. Put a lamp a little way behind the chairs. Switch on the lamp and shine it on the screen.

## What's going on?

Light from the lamp passes through the unpainted areas of the screen. But the puppet blocks the light, casting a shadow onto the screen. From the other side of the screen, the audience will see sharp outlines of the scenery and the shadows of the puppets.

The shadow will fall on the paper.

8. Sit or kneel behind the chairs and the screen. Hold your puppet by the straw, so that it's almost touching the paper.

9. Move your puppet up and down to make it perform. The audience will see the puppet's shadow on the screen.

The audience can see the shadow from this side.

For a link to a website where you can try some simple shadow show experiments in an online lab, go to **www.usborne-quicklinks.com**

# Light and color

Light may look white, but it's actually made up of red, orange, yellow, green, blue, indigo and violet. When water in raindrops reflects sunlight, it splits it into these seven colors and makes a rainbow. Here you can find out how to make rainbows and discover why sky looks blue and a sunset looks red.

These are rainbow patterns created by nail polish.

## Make rainbow paper

1. Half fill a big bowl with water. Add one drop of clear nail polish to the surface of the water. It will spread out.

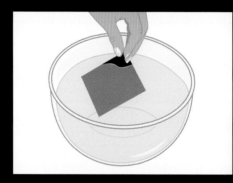

2. Dip a small piece of black paper into the water and lift it out. Let it dry. If you tilt the paper you will see rainbows.

### What's going on?

The nail polish forms a thin layer on the water. When the nail polish is transferred to the paper and light shines on it, the light is reflected by the layers of nail polish. This creates rainbow patterns.

## Rainbow reflected

You may need to prop up the mirror with a small stone to keep it in place.

1. Fill a tub with water. Lean a mirror at an angle at one end. Shine a flashlight onto the underwater part of the mirror.

2. Then hold a sheet of paper a little way behind the flashlight. Move it around until you see a rainbow on it.

### What's going on?

When a beam of light passes through water, the water makes the light bend. The different colors in light bend by different amounts, which makes them separate, making a rainbow. The mirror reflects the rainbow onto the paper.

For a link to a website where you can learn more about color, go to www.usborne-quicklinks.com

# Surprise spinner

1. Put a mug on a piece of white cardboard and draw around it. Then cut out the circle you have made.

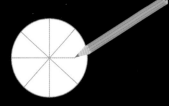

2. Using a ruler and a pencil, divide the circle, like this. There should be eight sections.

Felt-tip pens are most effective.

3. Fill one section of the circle in red, the next green and the next blue. Repeat this until all the sections are filled in.

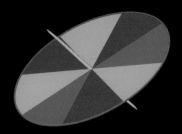

4. Using a thumbtack, make a hole in the middle of the circle and push a toothpick halfway through it.

5. Rest the toothpick inside a straw. Use your other hand to spin the circle as fast as you can. What do you see?

## What's going on?

When the spinner turns fast, your eyes see all three colors at the same time. Your brain can't separate them. So it combines them to make white, or white with a grayish tinge.

# Sky and sunset jar

1. Put half a teaspoon of milk into a jar. Then fill the jar with water to make a cloudy white mixture.

2. In a dark room, hold a flashlight to one side of the jar and shine it through the jar. The mixture will look blue.

The flashlight is behind the jar.

3. Now move the flashlight so that it is behind the jar and shines through at you. The milky water now looks red.

## What's going on?

The milk acts like air particles in the sky, which scatter different colors of light in different directions. This affects what colors we see, which is why sky sometimes looks blue, sometimes red. When light shines through the side of the jar, blue is the light that is most likely to be scattered and seen. When you hold the flashlight behind the jar, the light is scattered in a different way. You see red, the light that hasn't been scattered so much. This is similar to what happens in a sunset.

# Seeing things

When you watch a movie at the movie theatre, it appears to be one long moving image. But you're actually seeing lots of still pictures, each one slightly different from the next. Your eyes and brain translate these into a moving picture. These two experiments trick your brain in a similar way.

## Flip book

1. You need a small pad of paper, thin enough to see through for tracing. Draw a stick man on the last page.

You can see the first picture through the page.

2. Turn to the second to last page. Trace the outline of the man, but very slightly change the position of an arm or leg.

This stick man is kicking his leg.

3. Keep tracing the picture from the page below, but make small changes each time as if the man had moved a bit.

4. Draw at least 20 pictures like this. Then flip through them from back to front. Your stick man will appear to move.

In movies, the pictures are joined together in a long strip, so that they can quickly pass through the projector.

## What's going on?

As you flip the pages, your eyes and brain try to blend the pictures together, so the stick man seems to move. Movies need to show 24 pictures every second to make the image smooth enough to look as if it's really moving.

# Catch the birdie

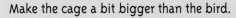
Make the cage a bit bigger than the bird.

You could also use yarn.

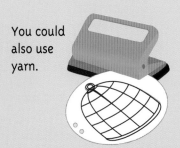

1. Make two circles on a piece of thin, white cardboard, by drawing around a mug twice. Then cut them out.

2. Draw a bird on one circle and a cage on the other. Turn the cage upside down. Glue them together back to back.

3. Use a hole puncher to make two holes on either side of the cage. Cut two pieces of thin string as long as your arm.

Keep swinging the disc until all the string is twisted.

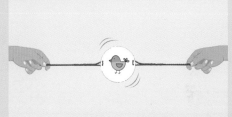

4. Thread a piece of string through one pair of holes, like this. Knot the ends. Do the same on the other side.

5. Hold the knots so that the circle hangs down. Flip the circle over and around until the string is twisted up tightly.

6. Now, with both hands, pull the string tight. This makes the circle spin around really fast. What do you see as it spins?

## What's going on?

As the circle spins, your eyes see one picture after the other. The pictures come around so fast that your brain can't separate them. Instead, it merges the two. So you see one picture — of the bird caught inside the cage.

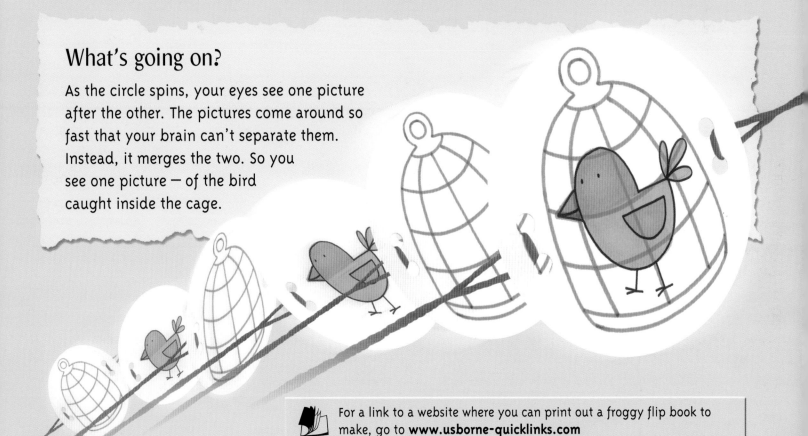

For a link to a website where you can print out a froggy flip book to make, go to **www.usborne-quicklinks.com**

17

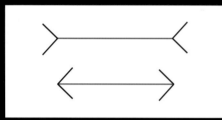

# Tricky pictures

Your brain uses short cuts so that it can process quickly all the things you see. Normally you don't notice these short cuts. But optical illusions such as the ones on these two pages trick your eyes and brain, so you see some very strange effects.

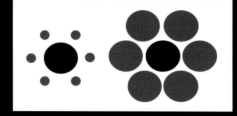

1. Which of the two red lines is longer? Measure them with a ruler to check whether you were right.

2. Which of the black circles is larger? Measure them with a ruler to find out if you were right.

## What's going on?

The red lines are the same length and the black circles are the same size. But the lines and circles around them trick your brain into thinking they're different sizes.

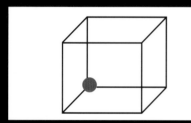

1. Is the red spot at the back or front of the box? Can you get it to switch places just by staring for a long time?

2. What do you see in the picture above? Do you see two people looking at each other — or an ornate vase?

3. Do you see four arrows pointing into the middle of the square, or four arrows pointing to the corners?

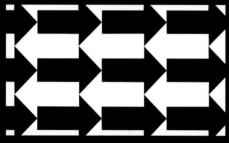

## What's going on?

Your brain can switch between two ways of looking at each of these pictures. This is because the pictures lack the extra details and shading that would normally help your brain figure out which view is the right one.

4. Which way are the arrows pointing? Do you see white arrows pointing left or black ones pointing right?

 For links to websites where you can see some more optical illusions, go to **www.usborne-quicklinks.com**

# Ghostly shapes

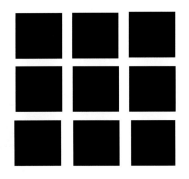

**1.** Can you see small gray squares in between the corners of the black squares of this pattern?

**2.** Do you see three circles each missing a wedge or a triangle floating in the middle of the circles?

## What's going on?

Strong patterns, like the black squares in the first example, can get blurred in your brain so you see ghostly gray squares instead. Your brain can also fill in details from simple clues. The missing wedges of the circles in the second picture become corners of a ghostly triangle.

# Straight or curved?

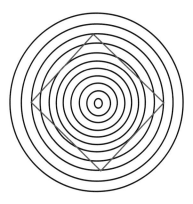

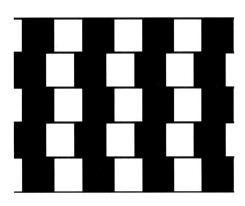

**1.** Is there a curvy or a straight-sided red diamond on top of these circles? Check with a ruler.

**2.** Are these tiles all the same size? Are the horizontal rows straight? Use a ruler to check whether you are right.

## What's going on?

Some patterns interfere with the way you see straight lines. The circular lines in the first example distort the straight-sided diamond, so that it appears curved. In the second example, the horizontal black line appears to be part of the black tiles. This makes the black tiles seem bigger than the white ones. So the lines appear slanted.

# Shades of gray

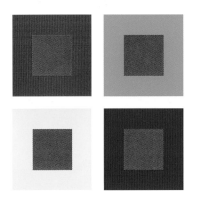

**1.** Which of these gray squares is darkest?

**2.** Now look at them without backgrounds. Which is darkest?

## What's going on?

The different backgrounds vary in brightness. This can trick your brain into thinking the gray squares are different shades. Without the backgrounds, it's much easier to tell the squares are the same.

# Sound vibrations

This is what sound vibrations look like on a computer screen.

All sounds are made by vibrations in the air. The vibrations reach your ear and make your eardrum vibrate. This is what makes you able to hear sounds. You can find out more by experimenting with vibrations on these two pages.

## Chiming fork

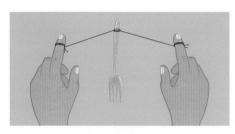

1. Cut a piece of thread as long as your arm. Tie the middle to the end of a fork. Wind the ends around your fingers.

2. First swing the fork so that it knocks gently against the edge of a table. You will hear a dull clink.

3. Now touch your index fingers to the flaps just in front of your ear holes and let the fork hang down.

4. Swing the fork so that it knocks gently against the table again. What do you hear this time?

Don't tie the thread too tightly around your fingers; it could restrict your blood supply.

### What's going on?

When the fork hits the table, it vibrates. This makes the air around it vibrate and you hear a dull clink. But it makes the thread vibrate too.

When you put your fingers near your ears, you bring the thread closer to the sound sensors in your ears. You can hear the vibrations much more clearly. They now make a clear chiming sound in your ear.

For links to a website where you can make a super sound cone, go to **www.usborne-quicklinks.com**

# Measure the volume

1. Stretch a piece of plastic food wrap as tightly as you can across the top of a large bowl.

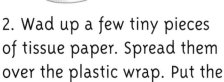

2. Wad up a few tiny pieces of tissue paper. Spread them over the plastic wrap. Put the bowl next to a speaker.

3. Switch on the stereo and play some music. Start with the volume low and gradually turn it up.

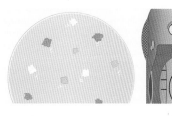

4. The tissue paper will start to move. Play different styles of music. What level of volume makes the paper move?

## What's going on?

The sound from the speakers makes the air vibrate. These vibrations get stronger as the music gets louder. Eventually they are strong enough to make the food wrap vibrate, so that the scraps of tissue paper move around.

Different styles of music have different speeds of vibration. Some will make the food wrap vibrate at lower volumes than others.

# Quacking duck vibrations

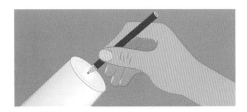

1. Make a hole in the bottom of a plastic cup using a thumbtack. Push a pencil into the hole to widen it.

2. Cut a piece of string as long as your arm. Then make a couple of knots in one end.

3. Thread the string through the hole, so that the knots rest on the outside of the bottom of the cup, like this.

4. Wet a paper towel. Hold the cup in one hand and drag the wet towel down the string with the other. What happens?

## What's going on?

As you drag the wet paper towel along the string, it makes the string vibrate. The vibrating string makes the cup vibrate too, which makes the sound louder. The vibrations are uneven, so they make an unmusical sound similar to a quacking duck.

The water in these bottles has food dye added to it, so that you can see the water levels more clearly.

# Highs and lows

Different types of vibrations make different types of sounds. Faster vibrations make higher sounds; slower ones make lower sounds. Try these experiments to discover how creating different vibrations can help you make musical notes.

## Bottle flute

1. Pour different amounts of water into a selection of glass bottles. Don't fill any to the top. Rest the neck of one bottle on your lower lip.

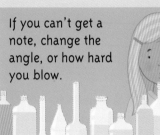

If you can't get a note, change the angle, or how hard you blow.

2. Blow gently across the top until you hear a note. Try all the bottles, comparing the notes they make. Arrange them from highest to lowest.

### What's going on?

Blowing makes the air inside the bottle vibrate. This produces a note. The notes change according to the amount of water and air in the bottle. The bigger the space between the water and the top of the bottle, the lower the note.

## Rubber band guitar

Paint the circle in the middle of the box.

1. Paint a circle in the bottom of a shoe box. Find two rubber bands the same lengths but different thicknesses.

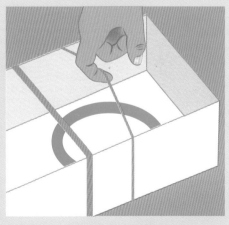

2. Stretch them over the box and pluck each one with your finger. The thinner one makes a higher note.

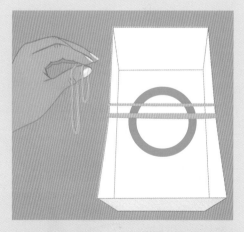

3. Now choose two bands that are the same thickness but different lengths. Which do you think will sound higher?

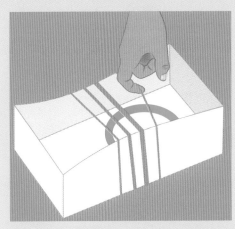

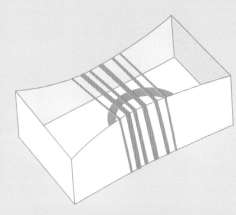

4. Stretch them over the box and pluck them. The shorter one makes a higher note. Were you right?

5. Stretch more rubber bands over the box. Pluck each one and arrange them in order, from the highest to the lowest.

This makes the neck of the guitar.

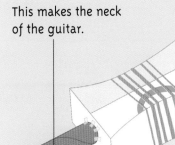

You could paint and decorate the guitar.

Rectangles

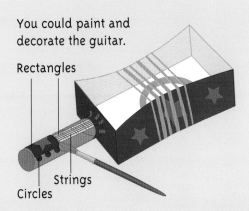

Strings

Circles

6. Find an inside tube from a roll of paper towels. Attach the tube to one end of your box with clear tape.

7. To make it look more like a guitar, you could paint on a rectangle and circles for a guitar head and draw strings.

## What's going on?

Thinner rubber bands vibrate more quickly, so they make higher notes. The thicker rubber bands vibrate more slowly and make lower notes. The more a rubber band is stretched, the faster it vibrates. So when stretched, shorter ones make higher notes than longer ones.

This is a six-stringed guitar, but you could make one with more strings.

 For a link to a website where you can make more instruments in an instrument lab, go to **www.usborne-quicklinks.com**

# Pushing and pulling

A force is a push or pull that makes an object do something. Without forces, nothing would ever start moving. In these experiments you can find out how forces make balloon rockets race along strings, and what makes it hard for forces to do their job.

## Fire a balloon rocket

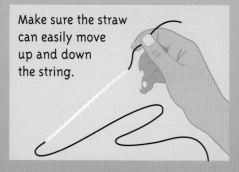

Make sure the straw can easily move up and down the string.

1. Cut a piece of string, about 10 feet long, and thread it through a straw. Tie one end to a dining chair.

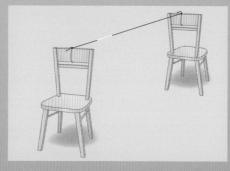

2. Tie the other end of the string to another chair. Pull the chairs apart to make the string tight.

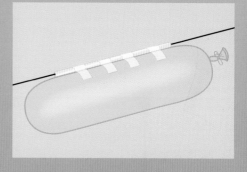

3. Blow up a balloon and hold the neck closed with a paperclip. Tape the balloon to the straw, like this.

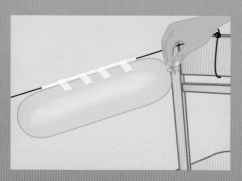

4. Push the balloon to one end of the string, with the neck facing a chair. Take the paperclip off. What happens?

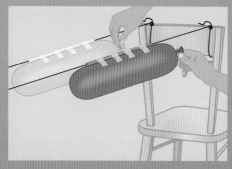

5. You could set up a second line and attach another balloon. Now you can race them with a friend.

## What's going on?

As the balloon deflates, it pushes the air inside out of the neck. The air flowing out pushes the balloon along the string in the opposite direction. Scientists describe this with a rule: every action has an equal and opposite reaction.

# Falling orange

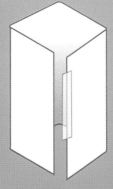

1. Cut a piece of cardboard 4 x 3 inches. Fold it into a rectangular column, like this, and tape it together.

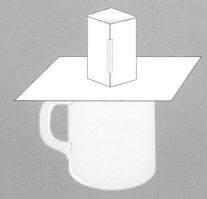

2. Lay a postcard on top of a mug. Put the cardboard column on top, so that it's over the middle of the mug.

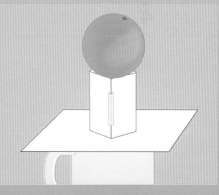

3. Carefully balance a small orange on top of the column so that the orange is directly above the mug.

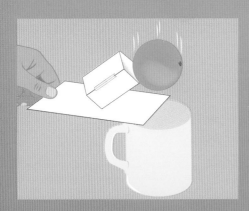

4. Pull the postcard away with a sharp tug. The column will fall to the side and the orange will drop into the mug.

## What's going on?

The column is light and easily moves sideways when you pull the postcard from underneath. But, as the orange is much heavier, it isn't moved so easily by the same pull. So the orange drops straight down into the mug. Scientists call this inertia. Inertia measures how hard it is for a force to move an object. The orange has high inertia, because it's heavier, and the column has low inertia.

For links to websites where you can design a roller coaster, and solve a forces fairground mystery, go to **www.usborne-quicklinks.com**

# Friction in action

Whenever an object moves across a surface, it is slowed down as the two surfaces grip each other. This gripping force is called friction. But some surfaces have more grip than others. Try these experiments to see friction at work.

## Spider slider

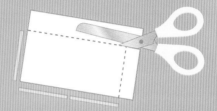

1. Snip the head off a used match. Then cut out a piece of cardboard two match lengths long and one wide.

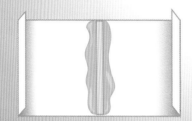

2. Press the match into some poster tack across the middle of the strip. Fold up a small piece at each end of the strip.

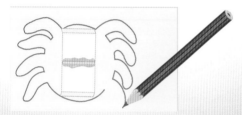

3. Draw a spider shape on bright paper. Make sure the spider is bigger than the strip of cardboard.

4. Cut out the spider shape. Make eyes and fangs from more paper. Glue them on to give the spider a face.

5. Glue the cardboard onto the back of the spider, like this. Then cut a piece of thread as long as your arm.

Remove the needle when the thread is through the holes.

6. Thread a needle with the thread and thread it through the middle of each fold of the cardboard strip.

7. Hold the thread tight between your hands, with one hand above the other. Then let the thread go slack. What happens?

### What's going on?

When the thread is held taut, it touches the match. This causes friction between the match and the thread, which is strong enough to stop the spider from moving down the thread. But, when you let the thread go slack, it no longer touches the match. This means there is less friction, so the spider slides down easily.

# Sliding race

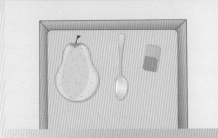

1. Line up some objects along one end of a smooth tray. Then lift that end and wedge a book under it.

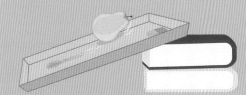

Do any of the objects slide?

2. Wedge more books under the tray, one by one. How high does the slope need to be before each thing moves?

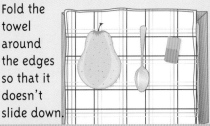

Fold the towel around the edges so that it doesn't slide down.

3. Now try it again, but this time put a towel on the tray first. Line up the objects on top of the towel.

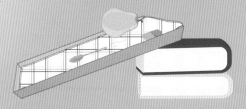

4. Wedge books underneath as before. Do things slide down at the same height? Do they slide in the same order?

## What's going on?

Rougher objects and surfaces create more friction than smooth ones. When you first start to tilt the tray, friction stops the objects from sliding. But, as the slope gets steeper, the smoother objects slide first. The towel makes the surface rougher, which increases friction. This means the slope has to be steeper before things slide. But they should slide in the same order. The weight of the objects can also affect how quickly they slide.

# Heating up

Don't move the other coin.

1. Lay two coins on a newspaper. Put a finger on one of them and move it very fast from side to side.

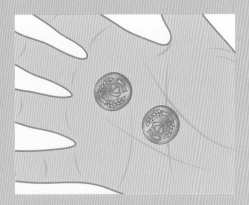

2. Stop moving it after half a minute. Lay both the coins on one hand. Which of the coins feels hotter?

## What's going on?

The coin you rub on the paper gets much hotter. This is because friction is caused by the movement of the coin on the rough surface of the paper. The energy that was going into moving the coin gets changed into heat energy during friction. You can feel the same effect when you rub your palms quickly together.

 For a link to a website where you can test friction with some wind-up toys, go to **www.usborne-quicklinks.com**

27

# Down to Earth

How high can you jump? How long can you stay in the air? No matter how hard you try, a force from the Earth called gravity will always pull you down. Try these experiments to find out more about how gravity and balance works.

## Balancing acrobats

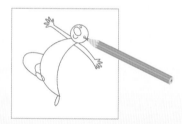

Make the ball about the size of a table tennis ball.

1. Roll modeling clay between your palms to make a smooth ball. Cut it in half with a knife to make a base.

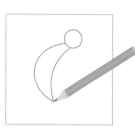

2. To make an acrobat, get a piece of plain cardboard the size of a postcard and draw a banana shape with a head.

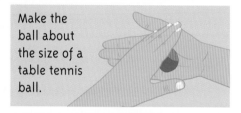

3. Then draw outstretched arms and another small banana shape for a leg. Add hands, feet and a face.

You can decorate your acrobat using felt-tip pens.

Tab

4. Draw a square tab under the foot. Then cut around the outside of the acrobat, including the tab.

Be careful not to squash the base out of shape.

5. Make a slit in the top of the base with a knife, then slot the tab into it. Now try to push your acrobat over.

## What's going on?

When you try to push over the acrobat, he pops back up. This is because the round base weighs more than the body. This uneven spread of weight affects how gravity pulls on the acrobat. The heavier the lower part of something is, the more easily it can stay upright.

# Balancing butterfly

The butterfly should be about the size of your hand.

1. Fold a piece of paper in half. Draw half a butterfly on it and cut it out. Unfold it. It will be symmetrical.

2. Draw around the butterfly on some cardboard and cut it out. Tape two coins on the tips of the wings, like this.

Try balancing it on your nose or the rim of a glass!

3. Now stick a pencil in some modeling clay. Try balancing the butterfly on the pencil. Where does it balance?

## What's going on?

The point where the butterfly balances is called its pivot point. If you change the weight on one side, or move the coins, the butterfly will balance at a different point. For the butterfly to balance, the weight of the coins multiplied by their distance from the pivot point must be the same on both sides.

You can make acrobats in lots of different poses.

For a link to a website where you can feel the pull of gravity when you try to land a spacecraft, go to www.usborne-quicklinks.com

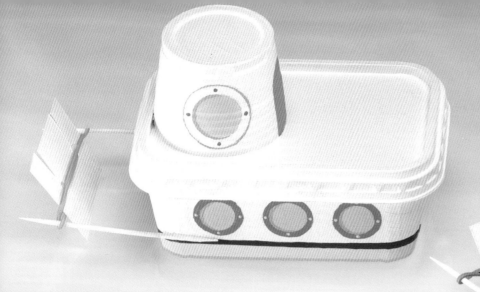

# Elastic energy

When you stretch a rubber band you use energy, which the stretched rubber band stores. When you let it go, the energy is released and the band snaps back. Here you can find out how to make use of this energy.

## Make a rubber band paddle boat

Keep the lid on.

1. Glue a toothpick halfway up one of the long sides of a small, empty margarine tub. It should stick out, like this.

2. Glue another toothpick in the same way to the other side. These will be the supports for your paddle.

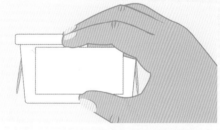

3. From another tub lid, cut out a piece the same shape as the end of the boat — but about ½ inch smaller all around.

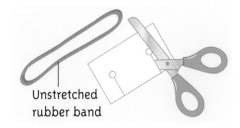

Unstretched rubber band

4. Make two holes in it with a hole puncher. Cut slits into the holes. Find a rubber band as wide as the end of the boat.

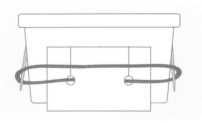

5. Slip the band through the slits into the holes. Then loop the band over the ends of the toothpicks.

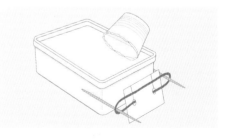

6. To make the captain's "bridge" for your boat, cut a plastic cup in half. Glue it to one end of the box lid.

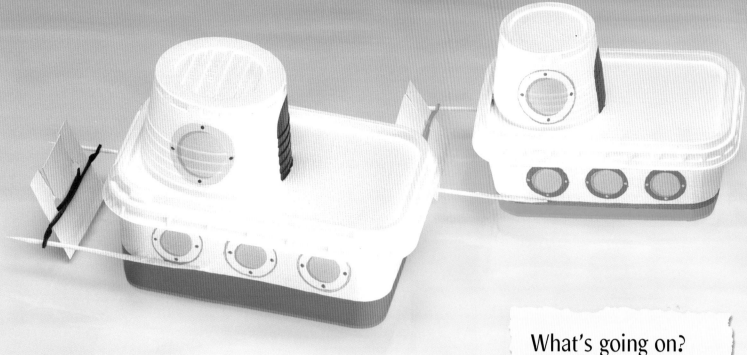

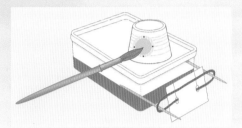

7. You could paint in details to your boat, like this. Then fill a bathtub or a sink with water and float the boat in it.

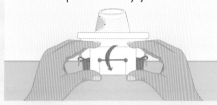

Twist the paddle away from the boat.

8. Wind the paddle until the rubber band is wound up. Then let it go. The boat should move through the water.

## What's going on?

As you wind up the rubber band, it stretches. When you let go, it unwinds and returns to its original length. The release of this stored energy turns the paddle. This is what pushes the boat through the water.

# Bouncy ball

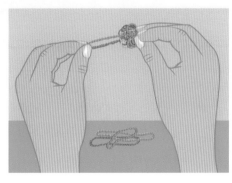

1. Squeeze four or five rubber bands together into a rough ball shape. Loop more bands over the ball, at different angles, to hold them together.

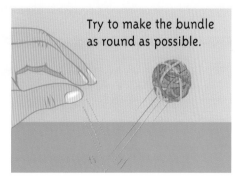

Try to make the bundle as round as possible.

2. When you have a ball about the size of a table tennis ball (15-20 bands), try bouncing it on the floor. How high does it bounce?

## What's going on?

When the ball hits the ground, the rubber is stretched a little, before it bounces back up and returns to its normal length. It is this stored energy, from the brief stretching, which provides the energy to bounce the ball back up into the air.

The world record for the biggest rubber band ball was 15ft across and made of six million bands!

 For a link to a website where you can try making other vehicles powered by rubber bands, go to **www.usborne-quicklinks.com**

# Stable structures

From tall skyscrapers to wide bridges, amazing structures need more than just strong materials. They need to be constructed in the right way too. Some shapes make stronger, more stable structures than others. Try these experiments to find out which shapes work best.

Try building a tall tower that can support a toy car.

## Tower challenge

Be careful! The spaghetti will snap easily.

The diagonals are about two thirds the length of a piece of spaghetti.

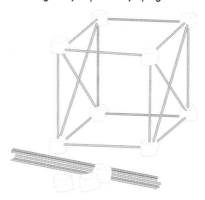

1. Use marshmallows and half lengths of uncooked spaghetti to build a cube like this. Does it feel stable?

2. Snap other pieces of spaghetti to make diagonals across each side of the cube. Does it feel more stable now?

3. Build the tallest tower you can from marshmallows and spaghetti. Put some cardboard on top and see what weight it will support.

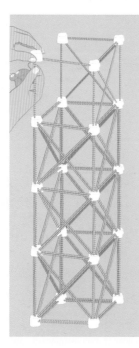

You could use several strands of spaghetti for each side to make it stronger.

Try joining several pyramid shapes together.

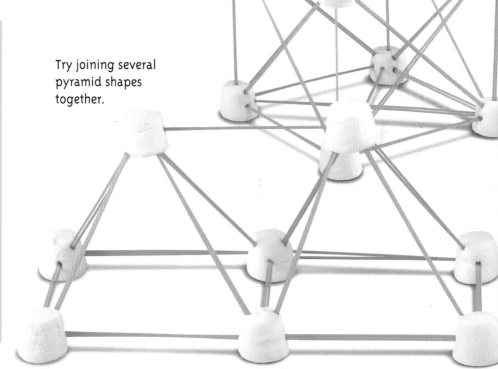

# Make a pyramid

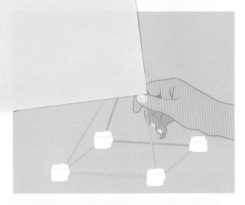

1. Make a square using half lengths of spaghetti and marshmallows. Add four more half lengths to make a pyramid.

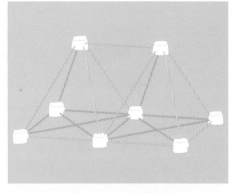

2. Add more spaghetti to extend your pyramid building like this. How stable does this shape feel?

# Building bridges

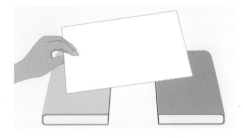

1. Put two heavy books of the same thickness, a handspan apart. Find a piece of cardboard the size of this book.

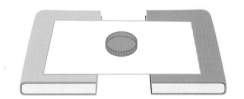

2. Lay the piece of cardboard on the books to make a flat bridge. Place a plastic lid from a jar on top.

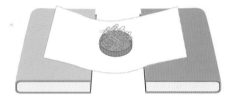

3. Add paperclips to the lid. The bridge will start to sag. How many paperclips does it take before it collapses?

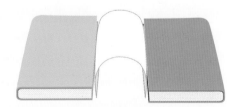

4. Put the lid to the side. Now make a second bridge with the cardboard. Bend it into an arch and put it between the books.

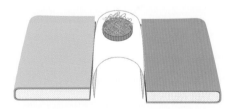

5. Balance the lid on top and add paperclips until the bridge collapses. How many does it take this time?

## What's going on?

The first bridge is flat and has nothing to support it at the sides. When more weight is added, it quickly collapses. When the cardboard bridge is arched, the weight is passed along the cardboard and into the heavy books at the sides. Because the weight is shared, the bridge can support a heavier load.

 For a link to a website where you can try some building challenges, go to **www.usborne-quicklinks.com**

# Under pressure

The air around you pushes against you all the time, and so does water when you're in it. Scientists call this pressure. When you squeeze air or water, its pressure increases. Try these experiments to see what happens to water and air under pressure.

The diver floats near the top of the bottle.

## Sinking diver

1. Find a piece of paper that will fit halfway around a large plastic bottle. Draw or paint an underwater scene on it and tape it around the bottle, so you can see it from the front.

2. Find a pen lid with a pocket clip, and attach a paperclip, like this. If there is a hole in the top of the lid, block it with a little poster tack.

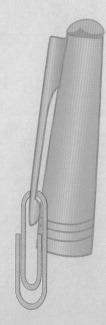

3. Cut out a diver shape from thin, colored plastic. Then press the diver on to the paperclip with poster tack.

The diver must be narrow enough to fit through the neck of the bottle.

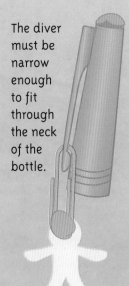

4. Put the diver in a tall glass of water. The model should float near the top. If it's too heavy and sinks, remove some of the poster tack.

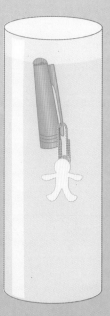

For a link to a website where you can try a pressure experiment with two straws and a drink, go to **www.usborne-quicklinks.com**

5. Fill the bottle with water. Then carefully lower the diver through the neck and screw the lid on.

6. Squeeze the sides of the bottle and the diver will sink. Then let go, and the diver will float up to the surface again.

The diver will move slowly at first, so watch carefully.

## What's going on?

An air bubble is trapped in the pen top when you drop the diver in. Squeezing the bottle pushes water up the top which squashes the air bubble and lets water in, making the diver sink. When you stop squeezing, the air bubble returns to normal size, pushing the water out. So the diver floats again.

# Keeping dry

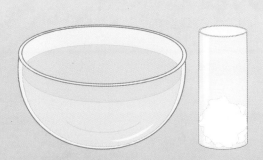

1. Fill a sink or large bowl with water. Then stuff a piece of paper into the bottom of a tall glass.

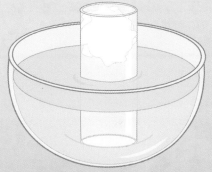

2. Plunge the glass into the water upside down. Then lift out the glass and check the paper. Is the paper wet?

## What's going on?

As the glass plunges into the water, the water pushes on the air inside the glass. The more the air is squashed, the more it pushes back against the water. Because they push at the same time, water doesn't get in and the paper stays dry.

# Sealed by air

1. Fill a glass to the brim with water. Lay a postcard on top of the glass, so that it totally covers the top.

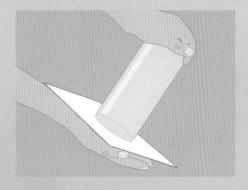

2. Holding the postcard against the glass, turn the glass upside down over a sink. Let go of the postcard.

## What's going on?

The postcard doesn't fall because air pushes against it and seals it to the rim of the glass. So the water stays inside the glass, instead of running out. The only way water can get out again is if you break the seal by removing the postcard.

# Taking flight

Air pressure — air pushing against you and everything around you all the time — is what keeps planes and birds in the sky. Here you can find out how air pressure makes wings work and makes a paper plane fly.

You could make your plane from any paper — even patterned paper.

## Make a paper wing

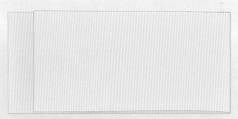

1. Cut a strip of cardboard about 6 x 6 inches. Fold it in two, so that one short end is ½ inch away from the other.

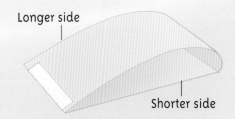

Longer side

Shorter side

2. Tape the two ends of the strip together. The longer piece will curve, making a wing shape.

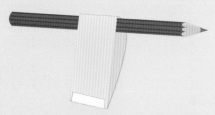

3. Put a pencil through the wing so that the wing hangs down, like this, with the flat side facing you.

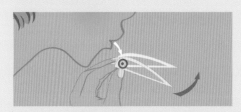

4. Blow down, just over the top of the pencil, onto the curved side. What happens to the wing?

## What's going on?

This wing shape is known as an airfoil. The air you blow over the curved side of the wing moves faster than the air under the straight side, because it has farther to go. This fast air is also lower in pressure. The slower-moving air underneath the wing has higher pressure. This high pressure is what pushes the wing up and keeps birds and planes in the air.

Air with low air pressure

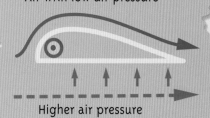

Higher air pressure

 For a link to a website where you can experiment with an online paper plane simulator, go to **www.usborne-quicklinks.com**

If you curl the tips of the wings down and throw the plane steeply upward, the plane will loop the loop.

# Make a paper plane

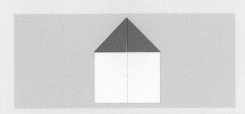

1. Fold a piece of letter size paper in half so the long sides meet. Open it out and fold the top corners to the crease.

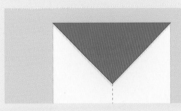

2. Fold down the whole triangle shape you've just made, so that the tip lines up with the crease in the middle.

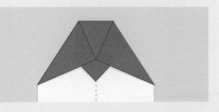

3. Then fold down the corners at the top so that they meet a little way above the tip of the triangle, like this.

4. Now fold up the tip of the triangle, so that it overlaps the folded-down flaps and holds them in place.

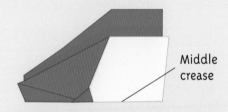

Middle crease

5. Turn the paper over. Then fold it in half down the middle crease and smooth the creases.

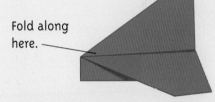

Fold along here.

6. To make wings, fold both sides down at the point shown here. Throw the plane to see how well it flies.

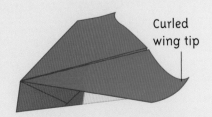

Curled wing tip

7. Curl the corner tips of the wings up or down around a pencil. How does this affect the plane's flight?

## What's going on?

As the fronts of the wings are thickest, they act in a similar way to the airfoil in the first experiment, helping the plane to fly. Curling the wing tips changes the air flow around the plane. Curling up the left tip will make the plane steer to the right and vice versa. Curling up both tips makes the plane climb. Curling them both down makes the plane dive.

# Magnetic attraction

Magnets pull on and attract some metals. Certain metals, such as iron, can act like magnets too. Explore magnetic attraction on these pages. You can make your own compass — which shows where north and south are — and find out how a compass works.

## Make a compass

1. Draw around a glass on a piece of thin paper. Cut out the circle. Then thread a big needle through it, like this.

2. Stroke the needle 20 times in the same direction with one end of a magnet. Lift the magnet between strokes.

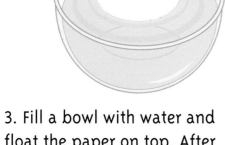

Be patient; it may take a moment before it moves.

3. Fill a bowl with water and float the paper on top. After a moment, it will slowly spin around and then stop.

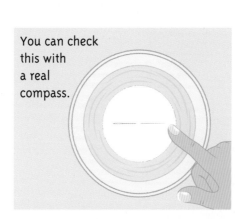

You can check this with a real compass.

4. If you turn the paper now, the needle will still spin back to point the same way. It will be facing north-south.

## What's going on?

A needle is made of steel, which contains particles of iron, jumbled up. But when you stroke a needle with a magnet, the iron particles become temporarily magnetized.

Jumbled iron particles in a needle

Inside the Earth there is so much iron that it acts like a giant magnet, giving the Earth a magnetic field. The magnetized needle lines up with the Earth's magnetic field. This makes it act like a compass, so it always turns to point north-south.

Ordered iron particles in a magnetized needle

For links to websites where you can try some online magnetism experiments, go to **www.usborne-quicklinks.com**

# Make a hovering butterfly

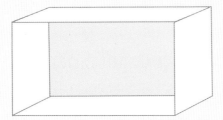

1. Lay a shoe box (without a lid) on its side. Then cut a piece of thread longer than the height of the box.

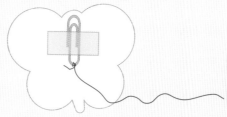

2. Tie a paperclip to one end of the thread. Cut a butterfly shape out of tissue paper and tape it to the paperclip.

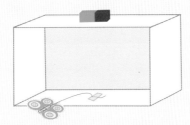

3. Hold the butterfly inside the box, almost touching the top. Pull the thread tight and tape it to the bottom.

4. Lay a magnet on top of the box, directly above the point where the thread is taped to the bottom.

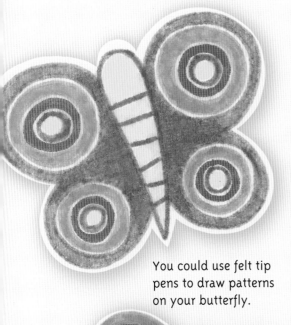

You could use felt tip pens to draw patterns on your butterfly.

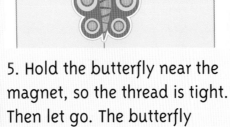

5. Hold the butterfly near the magnet, so the thread is tight. Then let go. The butterfly should hover by itself.

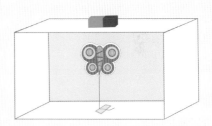

6. Try moving the butterfly farther away from the magnet, by shortening the thread. Does it still hover?

## What's going on?

Metal paperclips are made from steel which contains iron. The attraction between the magnet and the iron is strong enough for the magnet to pull on the paperclip, even without touching it. The thread stops the paperclip from being pulled onto the magnet. The stronger your magnet, the farther away you can move the paperclip and still make it hover.

# Static electricity

Some materials, such as plastic with wool — or synthetic fabrics such as acrylic — attract each other when you rub them together and can even cause a spark. This is called static electricity. Here you can see how to create static and investigate its effects.

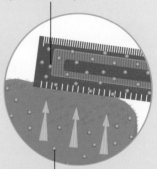

Static electricity enables this ruler to lift this paper snake.

## Be a snake charmer

1. Put a plate on a piece of tissue paper and draw around it. Cut out the circle. Draw a spiral snake inside it, like this.

2. To decorate your snake, draw a zigzag pattern and eyes with felt-tip pens. Then cut along the spiral.

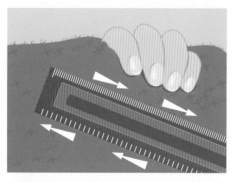

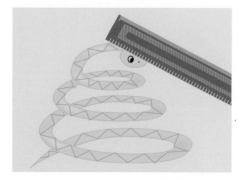

3. Rub a plastic ruler fairly hard and fast for half a minute with a scarf or sweater made of wool.

4. Then touch the snake's head with your ruler. Slowly lift the ruler. The snake should uncoil and rise up.

 For a link to a website where you can try some more static electricity activities, go to **www.usborne-quicklinks.com**

## What's going on?

When the wool is rubbed against the plastic ruler, it causes particles, that are too small to see, to pass from the wool to the ruler.

As you rub the ruler, it gains extra particles.

The particles are transferred from the wool.

These extra particles on the ruler cause a build-up of static electricity. The static pulls on the tissue paper. The tissue paper is so light that the static on the ruler is strong enough to lift it.

# Jumping pepper

You need to be able to see through the lid.

1. Find a shallow, plastic box and sprinkle a thin layer of ground pepper across the bottom of it. Put the lid on.

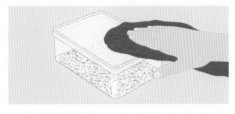

2. Rub the lid for about half a minute with a woolen scarf or sweater. Stop rubbing and watch the lid.

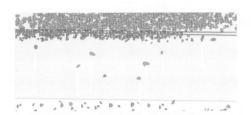

3. Specks of pepper will jump up and stick to the lid. You should see and hear them hitting the top.

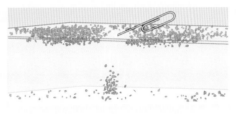

4. Unfold a metal paperclip. Touch the lid with one end of it. The pepper will move sideways or drop down.

## What's going on?

Rubbing the lid creates a build-up of static electricity, which attracts the pepper. When you touch the lid with the paperclip, the static is transferred to the metal. So the pepper drops down, or is pulled to other parts of the lid that still have static. The static travels through the metal paperclip, then your body and down to the Earth. So the paperclip doesn't get a build-up of its own.

# Testing static strength

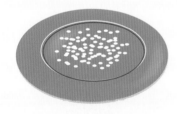

1. Make some paper dots using a hole puncher. You need enough to spread out across a small plate.

Rub in both directions.

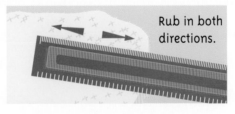

2. Rub a ruler 10 times across a woolen scarf or sweater. Press fairly hard as you stroke it.

Don't let the ruler touch the circles on the plate.

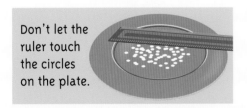

3. Hold the ruler just above the plate. Circles will jump up onto the ruler. Scrape them off and count them.

Tapping helps remove the static electricity.

4. Tap the ruler against a table. Then stroke the ruler against a different fabric and count the circles it picks up.

## What's going on?

Wool and similar synthetic fabrics are among the best materials for creating static electricity. It passes particles to the ruler very easily, so the ruler picks up lots of paper circles. But other fabrics, such as cotton, don't lose particles as easily. They make very little static, so the ruler doesn't pick up many circles. Tapping the ruler makes sure you're only testing the build-up of static from one fabric at a time.

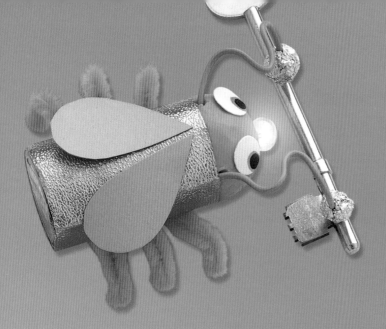

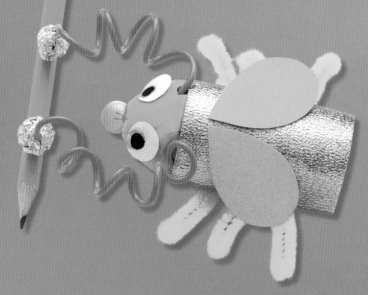

A key conducts electricity, so the bug's nose has lit up. But a pencil doesn't conduct electricity.

# Electrical bugs

Some materials allow electricity to flow through them. These are known as conductors. In this experiment, you can make a bug to test which materials conduct electricity. When the bug touches a conductor, its nose will light up.

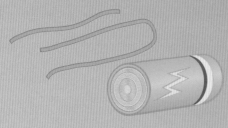

1. Cut a piece of plastic-coated electrical wire as long as a 1.5V battery. Cut another piece twice as long.

Gently press the scissors into the plastic all around the wire to make the cut.

2. Without cutting the metal wire, cut into the plastic a finger's width from each end and pull the plastic ends off.

Make sure the metal part of the wire touches the battery.

3. Tape one end of the long wire to the flat end of the battery. Tape the wire along the side of the battery.

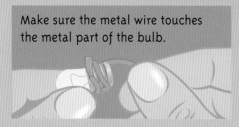

Make sure the metal wire touches the metal part of the bulb.

4. Wrap one end of the short wire around the metal base of a small flashlight bulb, like this.

The wire should come out of the side of the poster tack.

5. Hold the bulb base against the wire-free end of the battery. Surround it with poster tack to secure it.

6. Scrunch a palm-sized piece of kitchen foil tightly around each of the free ends of wire to make two balls.

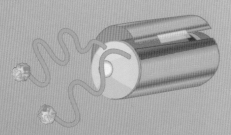

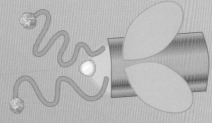

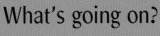

## What's going on?

Metal is a good conductor. When the foil balls touch metal, electricity flows from the battery and lights up the bulb. The electricity flows from the bulb along the wire, through the conductor and back to the other end of the battery through the other wire. This is called a circuit.

If the balls don't touch a conductor, there's no loop or circuit. Electricity can't flow, so the bulb doesn't light up.

7. Cut a piece of bright paper big enough to cover the battery. Wrap it around the battery and tape it on.

8. Draw two wing shapes on some thick paper. Cut them out and glue them onto your bug's back.

9. Cut two pipe cleaners in half. Tape three of the pieces across the bottom of the bug. Then bend them into legs.

10. Make eyes from circles of paper. Draw pupils in the middle. Glue them onto the poster tack near the bulb.

The arrows show the flow of electricity when the bug touches a conductor.

11. Now touch different things with both foil balls. If the objects are conductors, the bug's nose will light up.

Don't test electrical sockets or appliances. They could give you an electric shock.

 For a link to a website where you can test more materials to see if they conduct electricity, go to **www.usborne-quicklinks.com**

43

# Electromagnets

You can use electricity to create a magnet that you can switch on and off. This type of magnet is known as an electromagnet. Make your own electromagnet in this experiment and test how strong it is.

## Special equipment

You can get wire and a screwdriver from a hardware store and paper fasteners from an office supply store.

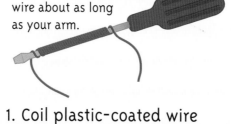

You will need a piece of wire about as long as your arm.

1. Coil plastic-coated wire around the metal part of a screwdriver. Leave a hand span of wire free at each end.

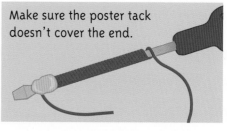

Make sure the poster tack doesn't cover the end.

2. Press a lump of poster tack near the end of the screwdriver, to stop the wire from coming off.

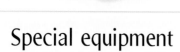

Gently press the scissors into the plastic all around the wire to make the cut.

3. Without cutting the metal wire, cut into the plastic a finger's width from each end and pull the plastic ends off.

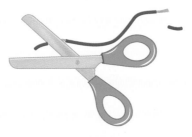

4. Cut another piece of wire a hand's width long. Strip the plastic off the ends in the same way as in step 3.

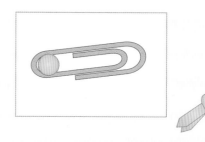

5. Lay a metal paperclip on a small piece of cardboard. Push a paper fastener through the paperclip and cardboard.

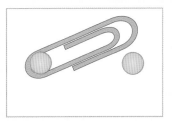

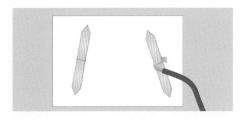

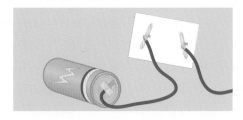

6. Push another fastener through the cardboard so that the paperclip can swing around and touch it.

7. Turn the cardboard over. Wrap one end of the long piece of wire around the back of one of the paper fasteners.

8. Wrap one end of the short wire around the back of the other paper fastener. Tape the other end to a 1.5V battery.

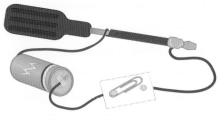

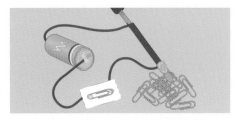

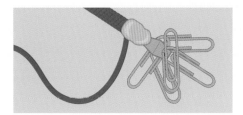

9. Make sure the paperclip is only touching one fastener. Then tape the free end of the long wire to the battery.

10. Move the paperclip so that it touches both fasteners. Now touch a pile of paperclips with the screwdriver's tip.

11. The screwdriver will act like a magnet and attract some of the paperclips. How many does it lift up?

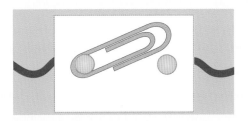

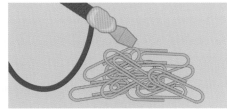

## What's going on?

12. Move the paperclip so that it only touches one fastener again. Leave it for a minute or two.

13. Touch the pile of paperclips with the tip of the screwdriver again. Does it pick up any this time?

The paperclip acts like a switch. When it touches both fasteners, electricity can flow from the battery through the wires. As electricity flows through, it creates a magnetic field. This magnetizes the metal of the screwdriver, so that it attracts paperclips.

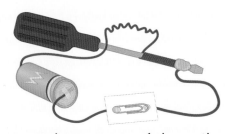

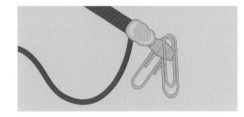

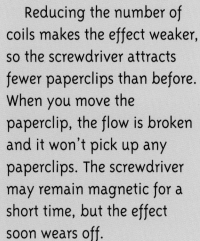

14. Unloop some of the coils of wire from the screwdriver. Then move the paperclip to touch both fasteners again.

15. Does the screwdriver lift the same number as it did in step 11? What happens if you remove even more coils?

Reducing the number of coils makes the effect weaker, so the screwdriver attracts fewer paperclips than before. When you move the paperclip, the flow is broken and it won't pick up any paperclips. The screwdriver may remain magnetic for a short time, but the effect soon wears off.

For a link to a website where you can build some simple electrical circuits online, go to **www.usborne-quicklinks.com**

# Freezing and melting

Everything exists as a solid, a liquid or a gas, but some things can change from one form to another. For example, water is a liquid, but when it freezes it turns into solid ice. You can find out about freezing and melting in these experiments.

## Making water grow

1. Fill a small plastic cup all the way to the top with water. Carefully stand it upright in the freezer. Try not to spill any water.

2. Leave the cup in the freezer overnight. Take it out when it has completely frozen. What has happened to the level of the water?

### What's going on?

Ice takes up more room than water. So, as the water freezes, it expands and takes up more space. This means that there is no room left in the cup, so it pushes up and freezes outside the cup. When the ice melts again, the water will go back to its original size.

## Slicing through ice

Carefully balance the wire, so the ice cube doesn't topple.

1. Open out a paperclip. Tape three heavy spoons to each end of the paperclip wire. Then balance an ice cube on top of a glass bottle.

2. Bend the wire so it's flat and balances on the ice. Put the bottle in the refrigerator for an hour. The wire should sink through the ice.

### What's going on?

The paperclip wire puts pressure on the ice. This melts the ice, so the wire moves down. As it does so, the water above it freezes again. This is similar to the way ice skates work on a rink. Their pressure temporarily melts the ice, so people are really skating over water.

# Making ice melt

1. Put your finger in the middle of an ice cube for 10 seconds. The pressure and heat will melt it a little.

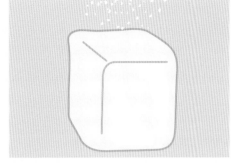

2. Sprinkle a pinch of salt on the middle of another ice cube. Leave it for a few minutes. What happens?

## What's going on?

Pressure and heat make ice melt faster. Another way to speed up melting is to use salt, because it makes ice melt at a lower temperature. That's why salt is mixed with the grit spread on icy roads.

# Fruity ice slush

1. Fill a mixing bowl with ice cubes. Sprinkle three tablespoons of salt on top and stir it in.

Don't let salty ice get in the glass.

2. Carefully place a glass upright in the middle of the ice. Half fill the glass with fruit juice.

If it's a hot day, you may need to add more ice.

3. Stir the juice every 10 minutes with a spoon. After about an hour and a half, the juice will become slushy.

4. Stir it every 5 minutes for another half hour until it becomes slush. Then you can eat it or leave it to freeze solid.

## What's going on?

Adding salt makes the ice melt at a lower temperature. In the bowl you get very cold salty ice and water. This mixture absorbs heat from the fruit juice, making the juice colder and colder. Eventually it will freeze solid, but stirring it breaks up the ice, so that it forms a slush instead of solid ice.

 For a link to a website where you can experiment with solids and liquids, go to **www.usborne-quicklinks.com**

47

You can make patterns like these in the experiment below.

# Surface tension

The surface of water and other liquids often behaves like a skin. This is because the tiny particles that make up the liquid pull on each other especially strongly at the surface. This is called surface tension. Watch it at work in these experiments.

## Breaking surface tension

1. Half fill a small bowl with water. Then sprinkle a thin layer of ground pepper on the surface.

2. Dip a toothpick in dishwashing liquid. Then touch the middle of the water with the tip of the toothpick.

3. As the dishwashing liquid touches the water, watch the grains of pepper. What happens to them?

Use several colors of food dye if you have them.

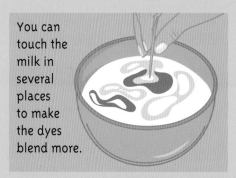

You can touch the milk in several places to make the dyes blend more.

### What's going on?

Dishwashing liquid reduces surface tension. This allows the particles of water at the surface to spread out more, starting where the dishwashing liquid was added. As they spread out, they push the pepper specks to the side. They also push the food dyes, so that they spread out and merge together, creating patterns.

4. Half fill another small bowl with milk. Then add two or three drops of food dye in different places.

5. Dip a toothpick in dishwashing liquid and touch the milk with it. What happens to the dyes as you do this?

 For a link to a website where you can try some more activities with surface tension, go to **www.usborne-quicklinks.com**

# Sink a paperclip

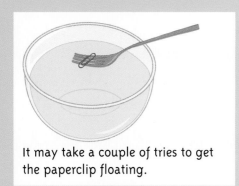

It may take a couple of tries to get the paperclip floating.

1. Half fill a bowl with water. Lower a paperclip onto the surface using the prongs of a fork. The paperclip will float.

2. Now mix a little dishwashing liquid with some water and pour it into the bowl. The paperclip will sink.

## What's going on?

The pull of the water particles at the surface — the surface tension — is just strong enough to hold up a paperclip. But adding dishwashing liquid reduces surface tension, so the water can't support the paperclip anymore and it sinks.

# Joining streams of water

Make the holes about a pencil's width apart.

1. Press a thumbtack into a large plastic bottle, to make three holes in a row near the bottom of the bottle.

2. Stand the bottle in a sink and fill it with water. The water will squirt out of the bottle, making three streams.

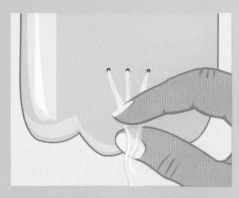

3. Pinch the three streams between your finger and thumb, an inch or so away from the bottle. Then let go.

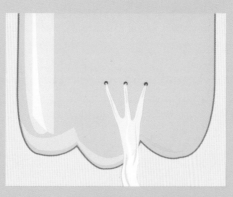

4. This makes the streams merge into one. If you drag a finger across the holes, the streams will separate again.

## What's going on?

When the streams are flowing separately, they are too far apart for the water particles in one stream to pull on particles in the other streams. But when you pinch the streams, the particles are brought close enough together to pull on each other. This pull holds them in one stream, even when you let go. Dragging your finger over the holes breaks the pull, and so they flow separately again.

Water particles in two separate streams

Two streams whose particles are pulling together

49

This is gloop made from cornstarch and water.

# Mixing

Some things mix together really well — sugar in tea, for example. But other things don't mix at all, or mix together with surprising results. You can investigate some mixtures in these experiments. They're a bit messy, so wear an apron and work somewhere you can easily wipe up spills.

## Making gloop

1. To make gloop, put two cups of cornstarch into a big bowl. Add a cup of water and a drop or two of food dye.

2. Mix the cornstarch, dye and water with your hands. It will take a few minutes to blend them all together.

3. Roll some of the mixture into a ball between your hands. What happens when you stop rolling?

4. Punch the mixture. How does it feel? Hold it up and let it dribble through your fingers. How does it feel now?

## What's going on?

Cornstarch is made of lots of long, stringy particles. They don't dissolve in water, but they do spread themselves out. This allows the gloop to act both like a solid and a liquid.

When you roll the mixture in your hands or apply pressure to it, the particles join together and the mixture feels solid. But if it is left to rest or is held up and allowed to dribble, the particles slide over each other and it feels like a liquid.

# Floating egg

1. Half fill a glass with water. Gently try to float a fresh egg in it. What happens to the egg? Take the egg out.

2. Now stir five teaspoons of salt into the water. What happens if you try to float the egg in the water now?

# Oily mixtures

1. Measure out three tablespoons of vinegar and three tablespoons of olive oil into a clean jar.

2. Notice how the oil floats in a layer on top of the vinegar. This is because the two liquids don't mix.

3. Now screw the lid on the jar tightly and shake the jar for about 30 seconds. How does the mixture change?

4. If you leave the new mixture for a few minutes, the liquids will separate and the layers reappear again.

5. You can use the mixture as a salad dressing. Add a pinch of salt and pepper and shake it again first.

 For links to websites where you can make a liquid sandwich and try other strange mixtures experiments, go to **www.usborne-quicklinks.com**

# Separating mixtures

Have you ever thought about what things are made of? Lots of everyday things are mixtures of other things that can be separated. These experiments show you how you can separate ink and cream, which are both mixtures of things.

Each of these strips was made using a different colored marker.

## Climbing ink

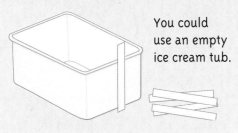

You could use an empty ice cream tub.

1. Cut some white blotting paper into strips slightly longer than the depth of a large plastic tub.

2. Make a dot with a different colored felt-tip pen a little way up each strip. Write the color at the top in pencil.

## Special equipment

You can buy blotting paper from most office supply stores.

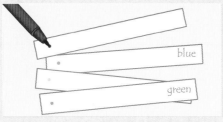

3. Pour just enough water into the tub to cover the bottom. Then tape a piece of string across the top of the tub.

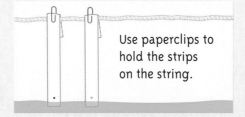

Use paperclips to hold the strips on the string.

4. Fold the strips over the string, so that the ends near the dots are in the water, but the dots aren't.

Blotting paper is very absorbent so water spreads quickly.

5. The paper will start to soak up water. Lift out the strips after ten minutes. What has happened to the dots?

## What's going on?

The ink in most felt-tip pens contains mixtures of different colors. Some colors dissolve more easily in water than others because of the chemicals they contain. These colors spread quickly up the paper. Other colors contain chemicals that don't like water. These colors stick to the paper to avoid the water. So they don't move up the paper as the water spreads.

Brown ink is made up of blue, yellow and pink. They separate out and become visible as the ink travels up the paper.

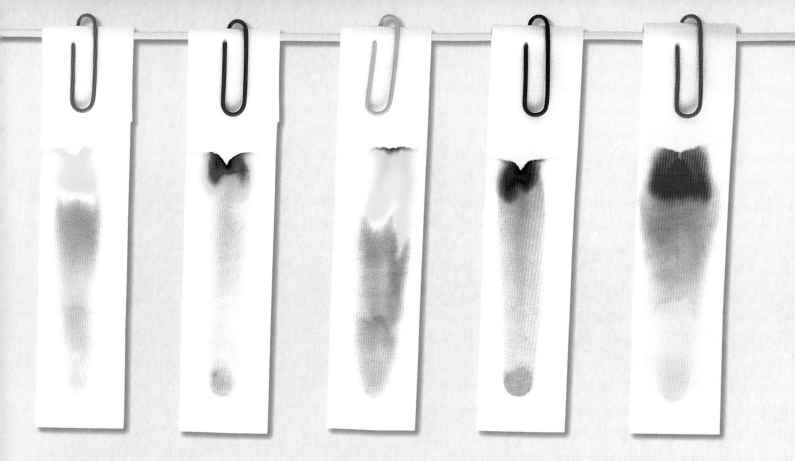

# Making butter

1. Half fill a clean jar with heavy cream. Add a pinch of salt for taste. Screw the lid on tightly and shake the jar.

It's tiring, so you may want a friend to help you shake!

2. Keep shaking it for 10-15 minutes. Eventually, it will separate into a lump of fat and a milky liquid.

3. Take out the lump and put it on a paper towel. Wrap the towel around it and squeeze out any excess liquid.

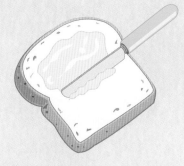

4. Now taste it. You have made butter. Put it in a dish, keep it in the refrigerator, and spread it on some bread.

## What's going on?

Cream is a mixture of tiny blobs of fat spread evenly through a milky liquid. When you shake the cream, the tiny blobs of fat bump into each other. The more you shake, the more they bump and join together. Eventually they turn into butter.

 For a link to a website where you can do an experiment to separate the colors in jellybeans, go to **www.usborne-quicklinks.com**

# Acids and alkalis

Acids and alkalis are types of chemicals. When they are very strong, they can be dangerous. But you can find weaker acids and alkalis in everyday things, such as lemon juice and baking soda. These experiments show you what they can do and how to tell them apart.

## Invisible ink

1. Pour a tablespoon of lemon juice onto a small plate. Dip your finger or a paintbrush in the juice and draw a picture on thin paper.

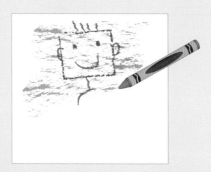

2. Leave the paper to dry. Your picture will be invisible, but if you rub the side of a wax crayon over the paper, the picture will appear.

### What's going on?

Lemon juice is an acid. The acid breaks down particles in the paper and weakens it, changing the surface of the paper. You can't see this change until you rub a crayon over it. Then the lemon juice parts stand out as darker than the rest.

## Smell eater

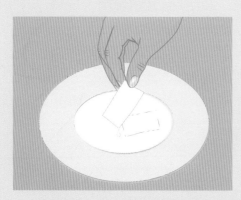

1. Cut two small pieces from a thick paper towel. Soak the pieces in milk and leave them to dry. How do they smell when they're dry?

2. Rub about a teaspoon of baking soda into both sides of one of the dry pieces. Now smell both pieces again. How do they compare?

### What's going on?

The piece of towel without baking soda rubbed into it smells sour. But the other piece of towel hardly smells at all. This is because baking soda is an alkali. Alkalis are often used in cleaning products and are able to cancel out or neutralize bad smells.

# Cabbage indicator paper

Chop the cabbage into small pieces.

Red cabbage juice stains. Don't get it on your clothes or furniture.

These will be your indicator paper strips.

1. Chop up half a red cabbage. Put it in a saucepan and cover it with water. Bring it to boil, then leave it to cool.

2. Rest a sieve on top of a large bowl and pour the cabbage mixture into it. Leave the drained liquid to cool.

3. Cut finger-sized strips from blotting paper or thick paper towels. Dip them in the liquid and leave them to dry.

4. Pour about half an inch of vinegar into a cup. Then pour half an inch of water into another cup.

5. In a third cup, add half a teaspoon of baking soda to about half an inch of water. Stir it well.

6. Dip a dry indicator strip into the cup of vinegar. What happens to the indicator paper?

7. Now test another strip in the water and one in the baking soda liquid. Does the same thing happen?

## What's going on?

The indicator papers change color when you mix them with an acid or an alkali. Acids always turn the paper red and alkalis always turn green. So you can use the paper as an acid and alkali detector. Vinegar is an acid and baking soda is an alkali. Water is neutral — it's neither acid nor alkali — and so doesn't make the paper change color. Try testing other things, such as carbonated drinks, tea or milk.

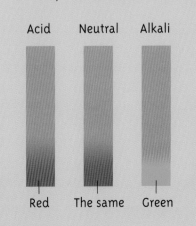

| Acid | Neutral | Alkali |
|------|---------|--------|
| Red | The same | Green |

 For a link to a website where you can visit an alien juice bar, go to **www.usborne-quicklinks.com**

# Foaming monster

When you mix acids and alkalis together, they make new chemicals. Some combinations, such as vinegar and baking soda, make a gas. Watch this chemical reaction at work by making this foaming monster.

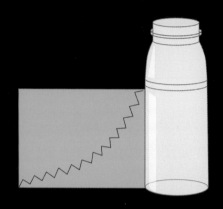

If you want to use your monster again, you can cover the paper parts in book covering film or tape to waterproof them.

1. Get a piece of thick paper, half the height of a small plastic bottle. Draw a monster's tail like this and cut it out.

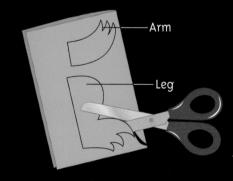

Arm

Leg

2. Fold another piece of thick paper in half. Draw an arm and a leg. Cut them out through both layers of paper.

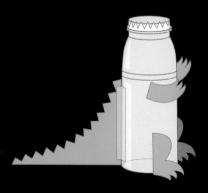

3. Tape the tail to one side of the bottle. On the other side, tape the legs to the bottom and the arms above them.

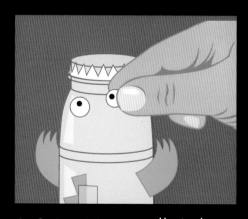

4. Cut out two small circles from white paper. Draw a dot on each one. Glue them above the tail to make eyes.

 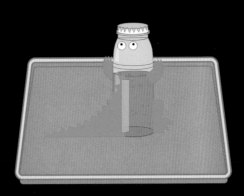 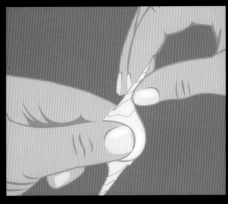

5. Half fill the bottle with vinegar. Add a good squirt of dishwashing liquid and a drop of food dye.

6. Gently swirl the bottle to mix the contents. Then place it in the middle of a large baking tray or dish.

7. Put a heaped teaspoon of baking soda in the middle of a square of tissue. Roll it up and twist the ends.

8. Drop the twisted tissue into the bottle. After a couple of minutes, foam will come out of the monster's mouth.

You can use different food dyes to change the color of the foam.

## What's going on?

When you mix vinegar and baking soda, it makes a gas called carbon dioxide. This forms bubbles in the vinegar. The bubbles of gas react with the dishwashing liquid to make foam. The whole combination reacts so much that foam pours out of the monster's mouth.

Baking soda was originally developed to help make bread and cakes rise. A similar reaction occurs in cake mix. The carbon dioxide bubbles make the mixture expand and rise.

 For a link to a website where you can do an experiment to make spaghetti dance, go to **www.usborne-quicklinks.com**

# Homemade paper

Paper is often made of thousands of tiny, long, thin strands of wood, squashed together. These strands, or fibers, can be separated out and reused to make new, homemade paper. In this experiment, you can try making your own recycled paper at home.

Wrap a rubber band around each end to hold the pantyhose in place.

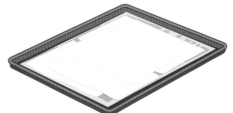

1. Bend a wire coathanger to make it into a square. Pull one leg of a pair of pantyhose over it to make a "screen."

2. Spread several layers of newspaper on a tray. Cover them with a layer or two of paper towels.

You can make paper with all kinds of different textures.

This paper has torn-up cotton balls mixed in, which make it stronger.

Add more water if the paper soaks it all up.

Rubbing the mixture between your fingers will help to break it up.

3. Tear thin scrap paper into small pieces. Put the pieces in a mixing bowl until you have about four cupfuls.

4. Add enough water to cover the paper. Leave it to soak for an hour. Then add a tablespoon of white glue.

5. Use your fingers to break up the paper into smaller pieces. After ten minutes or so it will be a thick mixture.

Make sure there are no gaps in the mixture.

6. To make the paper stronger, stir in torn-up cotton balls. Add food dye for color, or glitter for decoration.

7. Put the screen on top of the towels in the tray. Spoon the mixture onto it and spread it out in a thin layer.

8. Lay a plastic bag on top. Roll over it with a rolling pin, to even out the pulp and squeeze out the water.

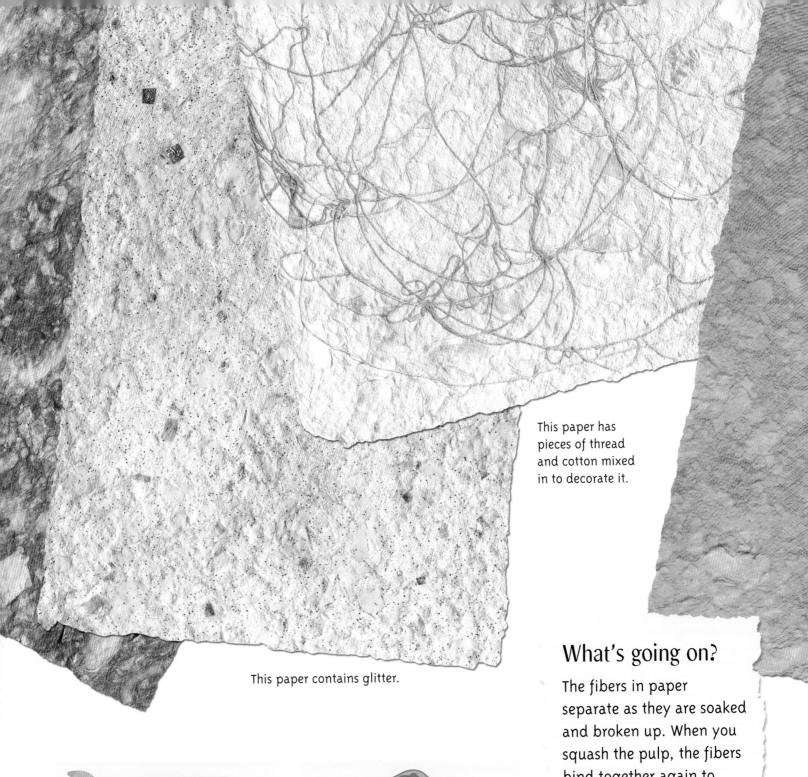

This paper has pieces of thread and cotton mixed in to decorate it.

This paper contains glitter.

## What's going on?

The fibers in paper separate as they are soaked and broken up. When you squash the pulp, the fibers bind together again to make new paper. Adding glue helps break down the pulp.

This is similar to what happens in paper recycling factories. There they add chemicals instead of glue to break down the paper, and have huge tubs to soak the paper in and heavy rollers to press and roll it.

Your paper will be fairly stiff.

9. Peel off the plastic bag and lift out the screen. Lay it on some fresh newspaper and paper towels. Leave it to dry.

10. After about three days the pulp should be dry. Peel it off the screen. You will have a piece of recycled paper.

 For a link to a website where you can find out how to make marbled paper, go to **www.usborne-quicklinks.com**

# Floating flowers

When paper gets wet, the thin wood fibers it's often made of absorb the water. This makes the paper swell and expand a little. With this experiment, you can see how this works and amaze your friends.

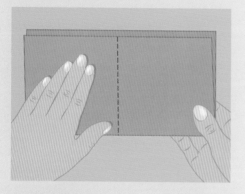

1. Cut out a square of paper about 6 x 6 inches. Fold it in half one way and then in half again.

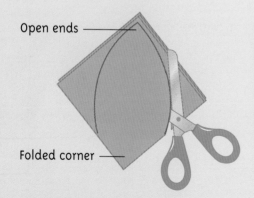

Open ends

Folded corner

2. Draw a petal shape outward from the folded corner. Cut around the shape to make the petals.

3. Open out the paper. Fold the tip of each petal to the middle point — the place where the creases cross.

4. To make a beetle to hide in the flower, draw an oval body on bright paper, add six little legs and cut it out.

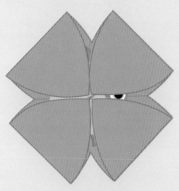

5. Cut out two wing shapes and two eyes in different-colored paper and glue them onto the body.

6. Put the beetle inside the flower. Fold down the petals. Fill a sink with water and lay it on top. What happens?

## What's going on?

As the fibers in the paper soak up water, they swell and the paper expands. As this happens there is a slight movement which makes the flower open up. Different types of paper soak up water at different speeds. Thin paper, such as newspaper, absorbs water very quickly. This makes the flower open up immediately. Other papers have thicker fibers, so it takes longer.

The paper lies flat when it's dry.

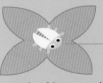

As the fibers swell, they push the paper outward and the petals open.

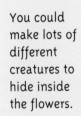

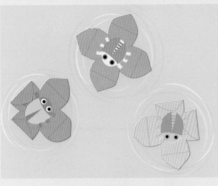

7. You could make more flowers from different types of paper and float them. Do some open faster than others?

You could make lots of different creatures to hide inside the flowers.

 For a link to a website where you can find more experiments to do with paper, go to **www.usborne-quicklinks.com**

# Chain reaction

Some chemicals found in food are made from long, curled-up chains of molecules. They're so tiny you can't see them, but they can have surprising effects when you mix them with other things. Find out more and do some cooking in these experiments.

## Milk shapes

These shapes have been decorated with glitter.

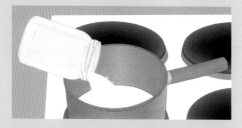

1. Half fill a jar with milk, then pour it into a saucepan. Gently warm the milk, but don't let it boil.

2. Turn off the heat. Add a drop of food dye and two tablespoons of vinegar. Stir the milk until lumps form.

Wrap a rubber band around the top to hold it in place.

You'll be left with little lumps.

Wax paper

Cookie cutter

3. Cut one foot off a pair of pantyhose. Put the toe inside the jar and fold the top over the sides, to make a strainer.

4. Pour the milk into the strainer and leave it for ten minutes. Squeeze out the rest of the milk into the jar.

5. Scoop the lumps out of the strainer and squeeze them into one lump. Press the lump into a cookie cutter.

The shape isn't edible!

6. Remove the cutter. Leave the shape on some paper. It will take a couple of days for it to dry.

## What's going on?

Milk contains chains known as casein, which are normally curled up and dissolved. When you add vinegar, they curl into a different shape and form solid plastic lumps instead.

 For a link to a website where you can make your own plastic products, go to **www.usborne-quicklinks.com**

# Making meringues

1. Cut a piece of baking parchment to fit inside a baking tray. Heat the oven to 225°F.

If the yolk breaks up, you will need to start again with a new egg.

2. Crack an egg on the edge of a bowl. Gently pull the shell apart and tip the white and yolk onto a saucer.

You can eat the meringues.

If you break a meringue in half, you can see the foamy texture.

You won't need the yolk.

Use an electric whisk if you can; it's quicker!

3. Hold a small cup over the yolk and tip the saucer so that the egg white dribbles into the bowl.

4. Beat the egg white. After about 15 minutes, it forms a thick foam and the whisk makes peaks when you lift it.

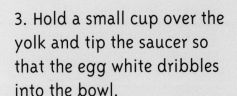

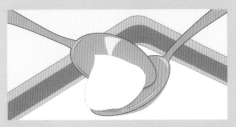

5. Gradually add ¼ cup finely granulated sugar. Whisk the mixture after adding each teaspoonful.

6. Take a heaped teaspoon of the mixture and slide it onto the baking parchment using another teaspoon.

## What's going on?

Egg white contains chains called albumin. Whisking whips air bubbles into the egg white. The albumin traps the bubbles, making a foam. When you bake it, the foam hardens into meringues.

Use oven mitts.

7. Do the same again, leaving gaps between each spoonful. Put the tray in the oven to bake for 45 minutes.

8. Turn off the oven and leave the meringues in for 15 more minutes. Then take them out and leave them to cool.

Before whisking, the albumin chains are quite tightly curled up.

Air bubble

After whisking, the chains uncurl and form a net that traps the bubbles.

# Creating crystals

You may think of crystals as expensive gems, but crystals can be found in lots of everyday things, including salt. Crystals are solids made up of particles arranged in a regular pattern. These experiments show you how you can make your own crystals.

Sugar crystals

Epsom salt crystals

## Candy crystals

1. Half fill a mug with hot water. Gradually stir in about two tablespoons of sugar until it is dissolved.

2. Cover two small plates with foil and pour two tablespoons of the liquid onto each plate.

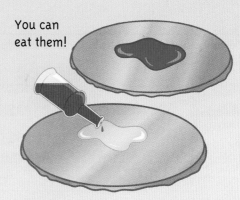

You can eat them!

3. Add a different drop of food dye to each. Leave them in a warm room. After three or four days, sugar crystals will form.

## Salt crystals

You can buy epsom salts from most pharmacies.

A dark plate will show the crystals more clearly. Don't eat them.

1. Half fill a mug with hot water. Gradually stir in about two tablespoons of epsom salts, until they are dissolved.

2. Pour two tablespoons of the liquid onto a small plate. Crystals will form after a couple of days.

### What's going on?

The water from the plates evaporates into the air and turns into water vapor — tiny water particles, spread so far apart that they are like a gas. Leaving the plates in a warm room speeds up the evaporation. Once the water has evaporated, only the crystals are left.

# Hanging crystals

When a layer forms at the bottom, it means no more will dissolve.

1. Fill two jars with hot water. Stir about six teaspoons of baking soda into each, until no more will dissolve.

2. Put the jars in a warm place where they won't get moved, with a small plate in between them.

The yarn should hang down but not touch the plate.

3. Cut a piece of yarn as long as your arm. Tie a paperclip to each end of it and place one end in each jar.

4. Leave the jars for a week. Crystals will grow along the yarn and hang down over the plate.

## What's going on?

The yarn soaks up the mixture. When the water evaporates, all that's left are baking soda crystals. The hanging crystals are formed when the mixture starts to drip from the yarn and evaporate. If you're lucky, you might even get crystals that drip onto the plate and form columns.

Be careful when pouring very hot water.

Baking soda crystals

The paperclip helps to weigh the yarn down.

For a link to a website where you can make some snow crystals, go to www.usborne-quicklinks.com

# Weather watch

Weather forecasters record changes in the weather to make predictions about what it will be like. In these experiments, you can record the direction of the wind, measure rainfall and air pressure, and make a model of an extreme weather condition.

This wind vane will show you which way the wind's blowing.

## Make a wind vane

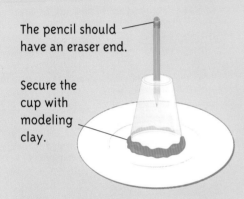

The pencil should have an eraser end.

Secure the cup with modeling clay.

1. Use a thumbtack to make a hole in the top of a plastic cup. Push a pencil through it. Secure the cup to a plate.

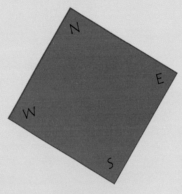

2. Cut out a square of colored cardboard and mark the corners, North, South, East and West, like this.

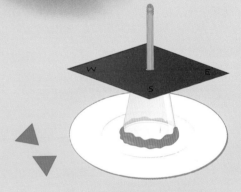

3. Cut a hole in the middle of the cardboard and push it over the pencil. Then cut two triangles from cardboard.

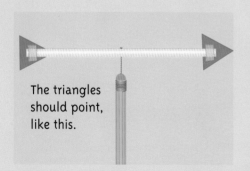

The triangles should point, like this.

4. Tape the triangles to the ends of a straw. Push a pin through the middle of the straw and then into the eraser.

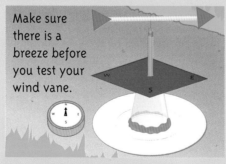

Make sure there is a breeze before you test your wind vane.

5. Put the vane outside and point it so that N matches North on a compass. Which way does the wind turn it?

## What's going on?

The wind blows on the wind vane and turns it until the arrows point in the direction the wind is coming from. You could make a chart to show which way the wind is coming from each day. The direction of the wind helps weather forecasters predict changes in the weather.

 For a link to a website where you can try predicting the weather, go to **www.usborne-quicklinks.com**

# Make a rain gauge

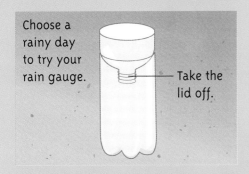

Choose a rainy day to try your rain gauge.

Take the lid off.

To stop it from blowing over, support it with stones, or sink it into the soil.

1. Cut the top third off a large plastic bottle. Upturn the top part and put it inside the bottom part. Stand it outside.

2. Use a ruler to measure how much rain has fallen each day. Empty the bottle every day and record your results.

## What's going on?

Measuring and recording the amount of rainfall is important because water is essential for life. Scientists compare rainfall in different countries and at different times of year, to see if, and how, the climate is changing.

# Measure air pressure

Secure the balloon with a rubber band.

The straw may move a little when the weather changes.

1. Cut the neck off a balloon and stretch the balloon tightly over a jar. Tape one end of a straw to the middle, like this.

2. Tape some cardboard behind the jar. Mark the end of the straw on it. Leave it for a day. Has the straw moved?

## What's going on?

The change may not be very marked. But if the straw tilts up, it means the air pressure is high because air pushes down on the balloon. Lower air pressure makes the air in the jar push up on the balloon, so the straw points down.

# Tornado in a jar

The technique for swirling the jar can take a couple of attempts.

1. Fill a jar three-quarters full of water. Add a teaspoon of dishwashing liquid and a teaspoon of vinegar.

2. Put the lid on and shake the jar. Now swirl it in a circular motion. A tornado-like shape will form in the jar.

## What's going on?

The liquids form a swirling motion, called a vortex. This looks very like a real tornado in a violent storm. A tornado is a swirling column of air, caused by changing temperatures and wind directions.

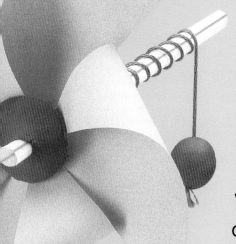

# Wind and water power

Most of the energy we use to produce electricity comes from coal, gas and oil. But, one day, the Earth's supplies of these fuels will run out. Before that happens, scientists will have to find alternative sources of energy, such as wind or water, that will always be available. These experiments show how wind and water can be used to provide energy.

## Wind power

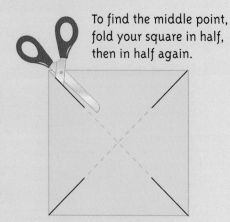

To find the middle point, fold your square in half, then in half again.

1. Cut out a square of bright paper, 4 x 4 inches. Cut halfway down from each corner to the middle, like this.

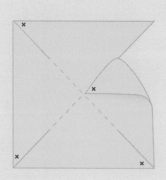

2. Fold the corners marked x to the middle and glue them down. The folds should curve and not lie flat.

3. Make a hole in the middle with a pencil and push a straw through. Secure it in position with poster tack.

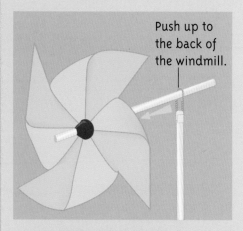

Push up to the back of the windmill.

4. Now tape a paperclip to a second straw, like this. Then push the windmill straw through the paperclip.

The poster tack should be the size of a pea.

5. Cut a piece of thread about the length of two straws. Stick a small lump of poster tack to one end.

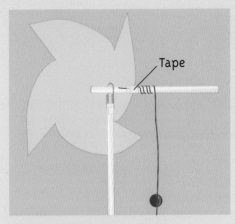

6. Tape the thread to the windmill straw. Wind the thread around it, leaving some hanging down.

7. Hold the other straw and blow to the side of the windmill. It will spin around, making the thread roll up.

## What's going on?

Your breath acts like wind and turns the windmill. This provides enough energy to pull up your small load of poster tack. Wind farms use much bigger windmills in the same way. The windmills turn machines and supply energy to generators to make electricity.

# Water power

Use a pencil to widen the holes.

1. Cut the top off a large plastic bottle. Use a drawing pin and a pencil to make six holes around the base.

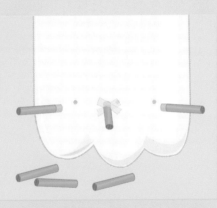

2. Cut a straw into six pieces about 2cm (1in) long and push them into the holes. Secure them with tape.

3. Make three holes at the top of the bottle and tie a piece of string through each hole. Then tie the strings to a fourth piece of string.

The strings should be about the same length.

4. Over the sink or outdoors, pour a jug of water into the bottle. As water pours out of the bottom, the bottle will spin around.

## What's going on?

The energy from the water pouring out of the holes, makes the bottle spin around. Falling water and its energy is used on a much larger scale at hydro-electric power stations. The water turns enormous wheels, called turbines. These drive machines called generators that produce electricity.

For a link to a website where you can play games and find out more about wind and water energy, go to **www.usborne-quicklinks.com**

69

# Sprouting seeds

When a seed sprouts, a tiny white root and a green shoot appear. These experiments help you find out what seeds need to grow into healthy, leafy plants. Beans, chickpeas and fruit pits are all types of seeds you can use.

## What grows best?

1. Take three plates and make a pile of ten paper towels on each one. Lay a cookie cutter on each pile.

2. Spoon water onto two of the plates to soak the towels. Write "dry" along one side of the pile without water.

3. Sprinkle cress seeds into each cutter. Hold down the cutter and spread the seeds to the edges with your finger.

4. Carefully remove the cutters. Put one watered plate in a cupboard, and the other two plates near a window.

5. Every day, add water around the seeds on the "wet" plates, but don't pour water over the seeds.

The cress plants should grow in the shape of the cutters.

6. After about a week, some of the seeds will have grown into plants. Which plates look the healthiest?

## What's going on?

The dry seeds don't grow at all, as seeds need water to sprout. But, once they've sprouted, they need light to make food, so the plants in the dark cupboard are yellow. The wet plate, by the window, grows successfully, as it has both water and light.

 For links to websites where you can help plants grow in an online lab and make an indoor garden, go to **www.usborne-quicklinks.com**

# Speedy shoots

You could use chickpeas or any dried beans, such as pinto beans or kidney beans.

1. Take two glasses of water. Put four lemon seeds in one and four beans in the other. Leave them to soak for a day.

You should be able to see the seeds through the jars.

3. Drain the seeds. Push the beans down the sides of one jar, and the lemon seeds down the sides of the other.

The beans should sprout after a few days. The lemon seeds will take longer.

5. Once the beans and seeds sprout, move the jars to a light place, such as a windowsill. Keep them wet.

The new pots give the plants space to grow.

6. They will grow roots. After a week or two, plant them root down in small pots of soil, and water regularly.

2. Stuff two jars with paper towels or napkins. Pour water into each one until the towels are soaked through.

4. Put the jars in a warm, dark cupboard. Check them each day and add water, if needed, to keep them wet.

## What's going on?

Putting the seeds in a dark cupboard encourages them to seek light and sprout. Once they've sprouted, light and water are essential for them to grow successfully. Beans and chickpeas sprout quickly. Most bean plants will grow and die in a year. Lemon trees live and grow for years, but their seeds may take several weeks before they sprout.

These chickpeas were planted 8 days ago.

This potted lemon plant first sprouted a year ago.

These beans were also planted 8 days ago.

71

# Soil science

Soil may not look all that interesting, but it's teeming with life and is vital to the growth of plants. Find out what's living under your feet and what your soil is made up of in these muddy experiments.

Woodlouse

Pill bugs

## Bug watch

It's easier if you make a hole for the scissors with a thumbtack first.

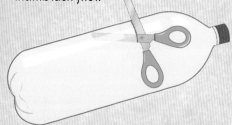

1. Cut the top third off a large plastic bottle. Take the lid off and put the top part upside down inside the bottom part.

2. Fill the top part with garden soil. Try to use soil with dead leaves on top, as it's a good place to find bugs.

3. Leave it under a lamp for two hours. Some bugs in the soil will burrow down and drop into the bottom part.

If no bugs fall through, try using soil from another area.

4. Do you recognize any of the bugs? Look at them with a magnifying glass. Then return them to the garden.

Place a small stone under the rim, so that insects can get inside.

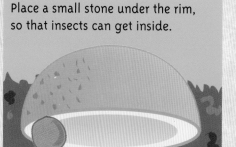

5. You could attract insects by scooping out half an orange. Leave it, upturned, on the soil overnight. Check it the next day.

## What's going on?

The bugs burrow down to hide from the heat and light. What you find in the soil will depend on where you live, where you get the soil from, and the time of year. Summer is probably the best time. You may find small varieties of bugs, beetles or worms in the bottle. The orange may attract bigger creatures, such as woodlice, slugs, snails and ants.

 For links to websites where you can play a bug hunt game and go on a soil safari, go to **www.usborne-quicklinks.com**

# What's in your soil?

1. Half fill a jar with soil from a garden. Then fill the jar almost to the top with water.

2. Screw the lid on tightly and shake the jar for about a minute. You will have a muddy, watery mixture.

3. Leave the jar for an hour. The mixture will settle into layers according to the weight of the soil particles.

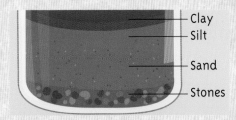

Clay
Silt
Sand
Stones

4. The layers will vary in thickness, depending on the soil you have. But it should settle in roughly this order.

5. You may see pieces of rotting plants and little bugs in the layers. Some of these may float to the top.

The best soil is a rich, even mixture of clay, sand and silt.

6. You will probably end up up with one thicker layer. This layer tells you what type of soil you have.

## What's going on?

The soils settle into layers according to how heavy they are. If your soil has even layers, it is described as loamy. This is generally the best kind, because it has a bit of everything. Clay soil contains lots of nutrients, which plants need to nourish their roots, but is not so good for drainage. Sandy soil drains very well, but has the fewest nutrients. Silt is finer and is in between clay and sand, but is more like clay.

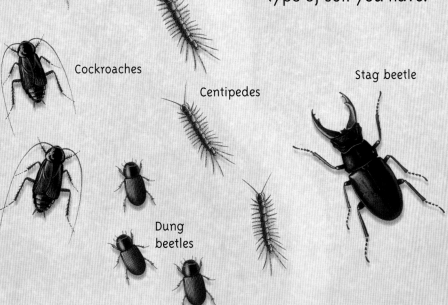

Millipede

Slug

Earwigs

Wasp beetles

Cockroaches

Centipedes

Stag beetle

Dung beetles

# Animal antics

Animals can behave in unusual and interesting ways. Even the smallest insects are smarter than you think. Try these experiments to see how ants organize themselves, and watch how worms' wiggling techniques help plants grow.

Ants work in a team and make organized trails like this.

## Team trail

1. First you need some ants. You may have to wait until the summer to find them on paths around your home.

If it ignores the fruit, move it in front of it again.

2. When you have found an ant, put a thin slice of fruit in front of it. It may eat some of it or carry bits away.

One ant may attract more ants, until they make a trail.

3. Check the fruit after an hour. Have other ants been attracted to it? If so, what are they doing?

4. When there are lots of ants, move the fruit to a new position, a little to the side. What do the ants do?

## What's going on?

Ants are one of the best examples of insect teamwork. They live together in big communities and help each other. So, if one ant finds a good source of food, it leads others there to eat it too. They follow each other by making long trails. The ants go back and forth, collecting food, bit by bit, to take back to their nest. When you move the fruit, the ants will still find it. But, instead of making a new direct route to the food, they will follow each other via their old trail route.

# Make a wormery

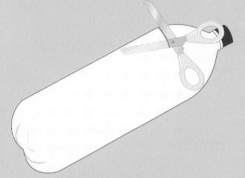

1. Make a hole with a thumbtack at the top of a large plastic bottle. Then cut the top off, like this.

If you don't have sand, try to find different colors of soil.

2. Fill the bottle with layers of soil and thinner layers of sand. Put dead leaves and four teaspoons of water on top.

You can often find worms under piles of dead leaves.

3. Dig around in some soil until you find two or three earthworms. Carefully put them into your bottle.

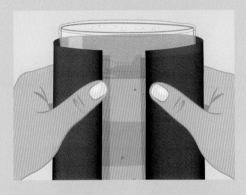

4. Cover the top of the bottle with food wrap and poke air holes in it with a pencil. Tape dark paper around the sides.

The paper has been taken off.

5. Add a couple of teaspoons of water each day, to keep the soil damp. After two weeks, take the paper off.

6. The worms will have mixed up the soil and made tunnels. Now return the soil and worms to their original home.

## What's going on?

Worms mix everything up as they make their burrows. The different-colored layers of soil and sand make it easier for you to see how they do this. It's great for gardens, as the mixing adds air to the soil and the burrows make channels for water. The worms may have pulled the dead leaves down to eat. This mixes nutrients into the soil. All this helps plants get everything they need from the soil to grow healthily.

 For links to websites where you can find out more about the fascinating worlds of ants and worms, go to **www.usborne-quicklinks.com**

# Butterfly search

Butterflies and moths have a fascinating, but short, life cycle. They start life on land and end their life with wings. Here you can see how a caterpillar changes into a butterfly, and find out how to attract butterflies to your garden. It's best to wait for late spring or summer to try these experiments.

## Caterpillar search

1. Use a sharp pencil to poke a few holes in the lid of a big, plastic ice-cream tub. Add a few pencil-sized twigs.

2. Look for a caterpillar on a leaf. Carefully put it, the leaf it's on and a few leaves from the same plant, in the box.

Don't put the box in direct sunlight.

The fresh leaves must be from the plant you found the caterpillar on.

3. Put the lid on and leave it somewhere warm. Check on your caterpillar every day and put in fresh leaves.

4. After about a couple of weeks the caterpillar should make a protective shell. This looks like a small brown case.

In a warm place, it should hatch between a week and ten days. If it hasn't hatched by then, carefully put it back outside.

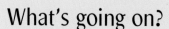

5. Now check the box twice a day. As soon as you see a butterfly or a moth, take the box outside and release it.

## What's going on?

The twigs provide a place for the caterpillar to spin its protective shell, called a cocoon or chrysalis. It forms this so that it can reform inside, and emerge as a butterfly or a moth. The plant that you found the caterpillar on is likely to be its source of food.

 For a link to a website where you can get ideas for carrying out more bug searches, go to **www.usborne-quicklinks.com**

# Make a butterfly feeder

Make knots at each end.

1. Use a thumbtack to make holes on opposite sides of the rim of a plastic cup. Then tie a piece of string through them.

2. Make a hole in the bottom of the cup using a thumbtack. Push a ballpoint pen into the hole to widen it.

3. Push a small cotton ball into the hole, so half is inside the cup and half is poking out of the bottom.

The petals should stick out from the base of the cup.

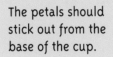

4. Cut out petal shapes from colorful plastic bags. Glue them to the cup, around the cotton ball, to make a flower.

## What's going on?

Sugary water is similar to nectar, the sweet liquid that butterflies drink from flowers. The bright petals attract butterflies to the feeder. They can then suck the sugary water as it soaks through the cotton ball.

A butterfly has a long tube called a proboscis to drink nectar from flowers.

5. Put nine tablespoons of water into a cup. Stir in a tablespoon of sugar. Pour the mixture into the plastic cup.

Don't stand too close or you may frighten the butterflies away.

6. Hang the feeder from a branch. Check on it from time to time during the day. Are any butterflies feeding?

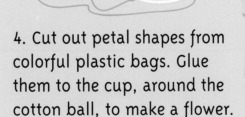

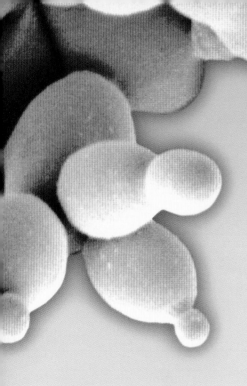

# Invisible creatures

All around us, there are thousands of tiny living things, called microbes. They are so small you can only see them through a microscope. Yeast is made up of microbes that make bread dough rise. See for yourself with this experiment and recipe.

## Yeast balloon

1. Mix two teaspoons of dried yeast with two tablespoons of warm water. Stir in a teaspoon of sugar. Pour the mixture into a small glass bottle.

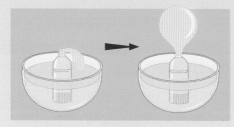

2. Stretch a balloon over the neck of the bottle. Stand the bottle in a bowl of warm water for 20-30 minutes. The balloon will inflate.

### What's going on?

The dried yeast doesn't react until you add the water and sugar. Then it starts feeding on the sugar. This produces bubbles of a gas called carbon dioxide, which blows up the balloon.

## Rising rolls

### You will need:

- ½ pint of warm water
- a teaspoon of sugar
- two teaspoons of dried yeast
- 1 ½ cups of all-purpose flour
- a pinch of salt
- a teaspoon of butter

1. Stir the warm water, a teaspoon of sugar and two teaspoons of dried yeast in a cup. Leave it for 10 minutes.

2. Put the flour, a large pinch of salt and a teaspoon of butter into a mixing bowl.

 For a link to a website where you can go on a quest to find bacteria in a cafeteria, go to **www.usborne-quicklinks.com**

3. Mix the butter, flour and salt by rubbing small amounts between your thumb and fingertips.

4. When the mixture looks like fine breadcrumbs, make a hole in the middle and pour in the yeast mixture.

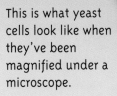

This is what yeast cells look like when they've been magnified under a microscope.

If it stays sticky, add a little more flour.

5. Mix everything together with your fingers. It will be sticky at first, but after a while it will form a dough.

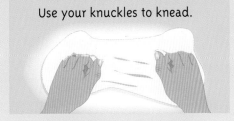

Use your knuckles to knead.

6. Sprinkle flour on a worktop. Then knead the dough by stretching and folding it for 10 minutes.

Gas from the yeast will make it rise and get bigger.

7. Return the dough to the bowl. Cover it with food wrap and leave it in a warm room for an hour and a half.

8. Take the dough out of the bowl and knead it for three more minutes. Then divide it into 12 equal pieces.

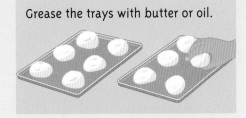

Grease the trays with butter or oil.

9. Shape the pieces into round rolls. Put six rolls onto each of two greased baking trays. Cover them with food wrap.

## What's going on?

When you add the yeast mixture to the flour it produces bubbles of carbon dioxide gas. The trapped bubbles of gas make the dough expand. When the bread is baked, the bubbles remain as tiny holes inside the bread.

10. Leave the covered trays for 30 minutes. Meanwhile, switch on the oven to heat up to 450°F.

Use oven mitts.

You can eat them.

11. Remove the food wrap. Then put the rolls into the oven for 12 minutes. Take them out and leave to cool.

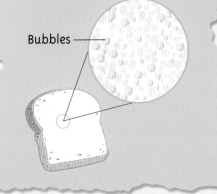

Bubbles

# Open wide

This is a photo of the surface of a tongue magnified many times. The larger red circles are taste buds.

Have you ever noticed how you can't always taste your food when you have a cold? Discover the importance of smell and saliva here, and find out if you think you taste different things better with different parts of your tongue.

## Tongue map

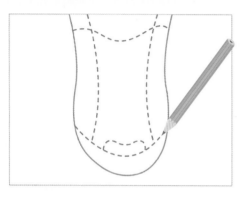

1. Draw a tongue on paper. Mark lines on it, like this, to divide up different areas of the tongue.

2. Take four cups. Put a different liquid in each one: lemon juice, cold black coffee, salty water and sugary water.

You could look in a mirror to help you do this.

3. Dip a cotton swab in the lemon juice and dab it on the areas of your tongue shown on the tongue map.

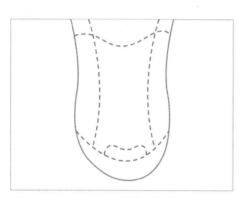

4. Does the lemon juice taste stronger in one part than another? Mark down the result on your tongue map.

5. Rinse your mouth with water. Repeat the test with the other liquids. Mark your results on the map.

## What's going on?

Some scientists think that your tongue works with your brain to recognize certain tastes on different areas of your tongue. They believe you taste bitter things at the back, sweet things at the front, sour things at the sides and salty things on the tip. But other scientists believe you can taste all flavors all over your tongue. What do you think?

For a link to a website where you can try out more experiments to test your senses, go to **www.usborne-quicklinks.com**

# Smelling power

1. Pour five different drinks into five plastic cups. You could use milk, water, juices and carbonated drinks.

2. Label what's in each cup. Then ask a friend to mix them up, so you don't know which is which.

3. Get your friend to pass you a drink at a time. Take a sip of each. Can you figure out which is which?

4. Mix up the cups again. Now take sips from them while holding your nose. Can you still identify them?

## What's going on?

Your nose is much more sensitive than your tongue. Without it, it's very difficult to identify tastes. You don't normally notice this because you smell and taste at the same time. Taking away your ability to smell shows how most of your sense of taste really comes from your sense of smell.

# Find the magic ingredient

1. Put a small amount of different dry foods on a plate. Try salt, sugar, a piece of cracker and a chip.

2. Stick your tongue out and dab it with a paper towel, until it's dry. Keep your tongue out.

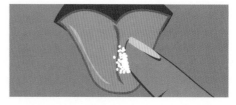

3. Use a clean finger to dab a little salt on your dry tongue. Can you taste it? Rinse your mouth with water.

4. Test the other foods, rinsing and drying your tongue in between. What has happened to your sense of taste?

## What's going on?

To be able to taste dry food, you need to mix it with saliva first. The taste can only be detected by the taste buds on your tongue after chemicals in the food have been dissolved in saliva. When you dry your tongue, there is no saliva, which makes it difficult for you to taste the dry foods.

# Test your reactions

Every time you move or touch something, hundreds of messages pass from your muscles and skin to your brain. The messages travel around your body through long fibers called nerves. Put your nerves to the test with these experiments.

This is what a nerve cell looks like when it's been magnified many times.

## Icy fingers

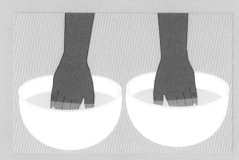

1. Sprinkle a few grains of rice on a small plate. Then put lots of ice cubes in a small bowl next to it.

2. Put your hand in the bowl of ice for 30 seconds. Dry your hand and try to pick up the rice. What happens?

### What's going on?

Your hand gets cold from being in the ice cubes. When your body is cold your skin is less sensitive, which dulls your sense of touch. This makes it harder to feel the grains of rice and pick them up.

## Hand thermometer

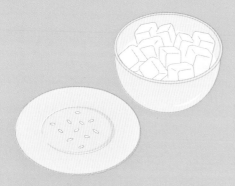

1. Fill one bowl with cold water and another bowl with lukewarm water. Put one hand in each bowl.

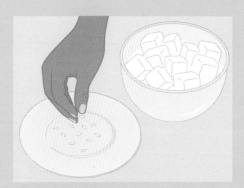

2. After a minute, move the hand in the cold water to the lukewarm water. Does the water feel hotter than before?

### What's going on?

The sensors in your skin that detect hot and cold get less sensitive as they adapt to different temperatures. The cold water makes the cold sensors in your hand less sensitive and the hot ones more sensitive. So, when you move your hand to the warm water, the water feels hotter than it is.

 For links to websites where you can test your reaction times, go to
**www.usborne-quicklinks.com**

# Quick catch

1. Ask a friend to hold the top of a long ruler. Curl your fingers and thumb around the zero mark at the bottom but don't touch the ruler.

2. Ask your friend to drop the ruler without warning you. Try to catch it between your fingers and thumb.

3. Check where your thumb ends up on the scale. This measurement shows how far the ruler has fallen.

4. Repeat the experiment a few more times. Can you catch the ruler any quicker with practice?

## What's going on?

You catch the ruler because a message travels from your eyes to your hand via your brain. There's a slight delay between the ruler dropping and you catching it while the message gets there. With practice, you can catch more quickly, but there's still a limit to how fast the messages can travel.

# Sensitive skin

1. Hold two pencils together, like this. Scribble with them to blunt the ends a little, so they aren't too sharp.

2. Keeping the pencils side by side, touch your fingertip with the points. Can you feel one point or two?

3. Touch your thigh with the pencil points. Can you feel both points? Move the points apart until you can.

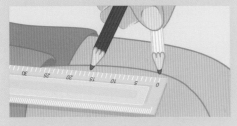

4. Measure the distance between the points. This distance is a measure of your skin's sensitivity.

## What's going on?

Some parts of your body, such as your fingertips, are very sensitive and have lots of touch sensors. This means you can feel two points even if they are close together. But other parts, such as your legs, don't need to be so sensitive, so the sensors are father apart. That's why the points of the pencil have to be farther apart on your leg before you can feel both of them.

# Heart and lungs

Your body needs oxygen to get energy. Your lungs take in oxygen when you breathe in air. The oxygen passes into your blood and your heart pumps the blood around your body. These experiments explore how you breathe and why your heart pounds when you run.

This is what red blood cells look like under a microscope.

## Make a plastic lung

1. Cut the base off a small plastic bottle. It's easier to cut if you pierce a hole with a thumbtack first.

You don't need this part.

2. Tie a knot in the neck of a balloon. Cut off the other end and stretch the balloon over the base of the bottle.

You don't need to blow up the balloon.

3. Push a straw into the neck of another balloon. Wrap a rubber band around it, to hold it in place.

Make sure the rubber band is tight but not crushing the straw.

4. Push the balloon end into the bottle, leaving the straw poking out of the top. Use poster tack to seal the top.

Press the poster tack firmly around the straw, but don't squash it.

5. Gently pull the knot of the balloon at the base. See how the balloon inside the bottle inflates a little.

Your **throat** and **windpipe** allow air into your lungs, like the straw.

Your **lungs** inflate with air, like the balloon inside the bottle.

Your **diaphragm** moves up and down to increase the space in your chest, like the outer balloon.

throat

windpipe

lung    lung

diaphragm

### What's going on?

Pulling on the outer balloon makes more space inside the bottle. Air rushes down the straw and into the balloon, inflating it to fill the space. When you let go, the balloon deflates again. This is similar to what happens in your lungs when you breathe.

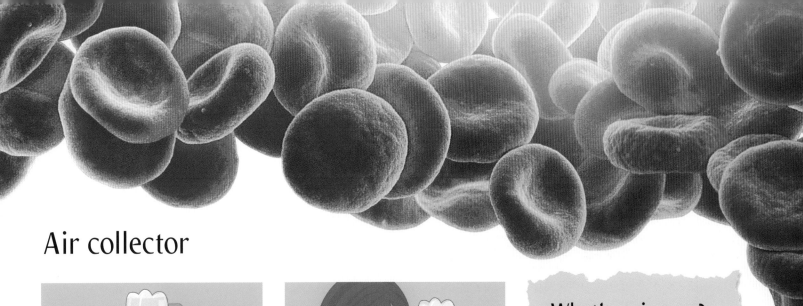

# Air collector

1. Fill a large plastic bottle with water and put the lid on. Then hold the bottle upside down in a big bowl of water and take off the lid.

2. Push a bendy straw into the neck of the bottle. Holding the bottle upright, breathe in, then gently breathe out fully into the straw. What happens?

## What's going on?

As you blow out, the air from your lungs collects at the top of the bottle. The more you breathe out, the more air fills the bottle and the more water drains out. This shows you how much air there is in your lungs.

# Rate your pulse

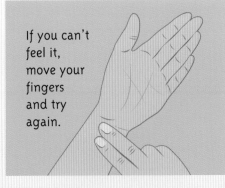

If you can't feel it, move your fingers and try again.

1. Rest for 10 minutes.

2. Then press your fingers on your wrist, just below the base of your thumb. You should feel a beat. This is called taking your pulse.

3. Using a watch, count how many beats you have in one minute. This is your pulse's resting rate.

4. Then run in place for five minutes.

5. Now sit down and take your pulse again for one minute.

6. Wait five minutes and take your pulse again.

7. Take it every five minutes until it returns to your resting rate.

## What's going on?

As your heart pumps blood around your body, your heart beats. You can feel these beats when you take your pulse. You breathe faster when you're running, to get more oxygen into your lungs and blood. This makes your heart and pulse beat faster, to get the oxygen you need more quickly to your muscles to make more energy.

 For a link to a website where you can see what makes your heart beat faster, go to **www.usborne-quicklinks.com**

# All in the mind

It's often difficult to remember people's names or long lists of things. Making links between pieces of information can help you to remember. Here, you can see how your brain is influenced by words and pictures, and practice tricks to improve your memory.

## Words and pictures

1. Look at the pictures on the right and say the name of each animal out loud. Do not read the word.

dog    bee    mouse    cat
bat    frog    squirrel    dinosaur
goldfish    seal    cockerel    elephant
dolphin    lion    crocodile    butterfly

crocodile    bat    cockerel    frog
dolphin    mouse    lion    cat
dinosaur    goldfish    butterfly    bee
elephant    seal    squirrel    dog

2. Now look at the pictures on the left. Say the names of the animals out loud. Is it harder?

## What's going on?

The first pictures are easy to name because the word below them is the same as the picture. The second group of pictures is harder because of the interference of different information. Most people can read quicker than they can name a picture. The different animal word below the picture confuses your brain.

For a link to a website where you can play some more mind games and try some "droodles," go to **www.usborne-quicklinks.com**

86

# Memory master

1. Study the list of objects below for two minutes. Close the book and write down as many as you can remember.

| | |
|---|---|
| saucepan | kitten |
| comb | book |
| crayon | ruler |
| sock | apple |
| lightbulb | umbrella |
| car | bed |

2. Now imagine going through your house putting the objects from the list in different places.

3. The more unlikely or funny the place, the easier it will be to remember the names of the objects.

4. Use this trick to try to remember the objects in the new list on the right. Do you remember more than before?

| | |
|---|---|
| goldfish | plant |
| scissors | shoe |
| chair | doll |
| lamp | calendar |
| banana | robot |
| key | hat |

## What's going on?

The imaginary trip around your house helps your brain build links between the objects to help you remember them. Unexpected or funny things, such as a goldfish in the toilet, make the list even easier to remember.

# Name that face

Choose people you don't know.

1. Cut out eight different faces from old magazines. Glue them onto pieces of cardboard and turn them face down.

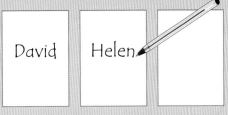

David | Helen

2. Write a name on the back of each card. Read out the name and then look at the face. Try to learn the names.

3. Mix up the cards. Then go through them just looking at the faces. Can you remember the names?

| David | Helen | Louie |
|---|---|---|
| painter | dancer | swimmer |

4. Now add a different hobby below each name. Mix them up and try to remember them. Is it easier?

## What's going on?

It's hard to remember names by themselves, as there are no other clues to help your memory. But adding extra information, such as hobbies, helps your brain form links so that you remember the name. Going through the names a second time also helps you remember.

# Family ties

All living things are made of cells. Each cell contains genes that are made of the chemical DNA. Genes carry the information that decides the characteristics of each living thing. DNA is invisible because it's too small to see, but in these experiments you can see it and discover what you've inherited from your family.

This is a model of part of a molecule of DNA. It looks a bit like a twisted ladder.

## Seeing DNA

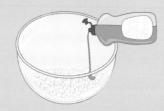

1. Finely chop an onion and put the pieces in a bowl. Mix in enough dishwashing liquid to coat but not cover them.

2. Add half a teaspoon of salt and two tablespoons of water. Stir gently to avoid making foam or bubbles.

## Special equipment

You can request denatured alcohol from pharmacies.

3. Leave the mixture for ten minutes. Then stir it again and use a sieve to strain off the liquid into another bowl.

4. Pour the liquid into a glass jar. Use a spoon to scrape off any foam or bubbles from the surface.

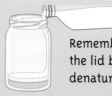

Remember to put the lid back on the denatured alcohol.

5. Pour denatured alcohol gently down the inside of the jar. The alcohol will form a separate layer. Don't stir it.

6. After about 20 minutes, a stringy white substance will appear in the top layer. This is the onion's DNA.

## What's going on?

The salt and dishwashing liquid help to break down the onion's cells, releasing the DNA. DNA doesn't dissolve in alcohol-based liquids such as denatured alcohol. So it appears as the solid, white strands that you see floating in the denatured alcohol on top of the dishwashing liquid.

# Make a family tree

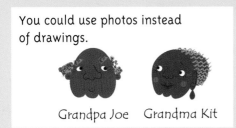

You could use photos instead of drawings.

Grandpa Joe    Grandma Kit

1. To make a family tree, start by drawing the faces and writing in the names of your grandparents.

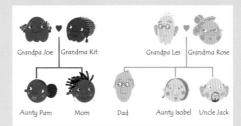

Grandpa Joe  Grandma Kit    Grandpa Les  Grandma Rose
Aunty Pam  Mom    Dad    Aunty Isobel  Uncle Jack

2. Underneath, draw lines like this to show your grandparents' children: your mom, dad, aunts and uncles.

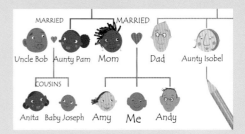

MARRIED    MARRIED
Uncle Bob  Aunty Pam  Mom    Dad    Aunty Isobel
COUSINS
Anita  Baby Joseph  Amy  Me  Andy

3. Then add who they married. Underneath, name their children: you, your sisters, brothers and cousins.

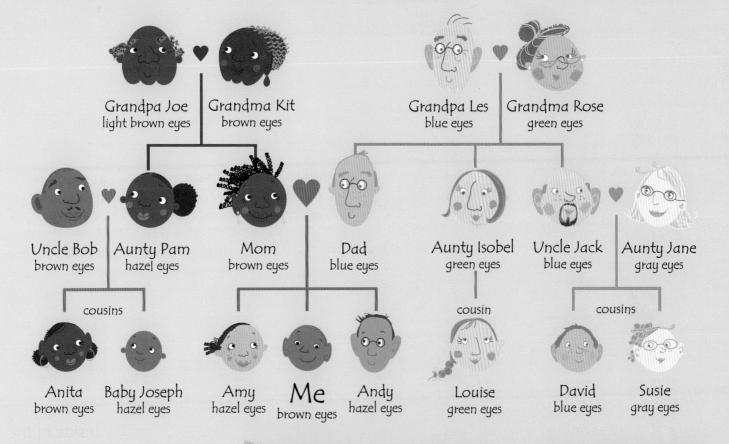

Grandpa Joe
light brown eyes

Grandma Kit
brown eyes

Grandpa Les
blue eyes

Grandma Rose
green eyes

Uncle Bob
brown eyes

Aunty Pam
hazel eyes

Mom
brown eyes

Dad
blue eyes

Aunty Isobel
green eyes

Uncle Jack
blue eyes

Aunty Jane
gray eyes

cousins

cousin

cousins

Anita
brown eyes

Baby Joseph
hazel eyes

Amy
hazel eyes

Me
brown eyes

Andy
hazel eyes

Louise
green eyes

David
blue eyes

Susie
gray eyes

4. Find out who can roll their tongue or wiggle their ears. Write it on the tree. Do you see any patterns?

5. Now add more detail to your tree. You could add eye color, height and nose or mouth shape.

## What's going on?

The ability to roll your tongue or wiggle your ears is only controlled by a few genes. This means these characteristics are likely to be inherited from your parents. Other characteristics, such as eye color or height are more complex. You could have inherited those characteristics from a relative further back in the family line.

 For links to websites where you can find out more about genes and make a DNA chain out of candy, go to **www.usborne-quicklinks.com**

# Doing your own experiments

Now that you've done the experiments and investigations in this book, you could try some scientific research of your own. Here you can find out how to plan, carry out and record your own experiment. You might be able to use your ideas and what you find out for a school science project.

## What topic?

Choosing what you want to investigate is often the hardest part of creating your own experiment. Think about what you're interested in. Try surfing the Internet or looking through books for ideas.

### My Experiment

**Question:**

What kind of ball bounces the highest?

**Answer:**

I think a tennis ball will bounce the highest!

**What I will need:**

tennis ball
soccer ball
golf ball
ruler
pencil

### Safety Tips

1. Never experiment with mains electricity or get water on electrical appliances or sockets.

2. Don't look directly at the Sun.

3. Be very careful when heating things, or using the oven.

4. Don't tie string or thread too tightly around any part of your body. It may restrict your blood supply.

5. Put the lids back on bottles when you've finished using them.

6. Don't pour anything into a bottle that's had something else in it, without rinsing it out first and relabeling it.

7. Wash your hands after handling soil, cleaning materials or chemicals.

## Question

Write down the question that sparked your idea. This will help you to be clear about what you want to find out.

## Answer

What do you think might be the answer to your question? This is called a prediction.

## Planning

Make a list of things you'll need to carry out your experiment. You may have to find or buy some of them.

## Recording

How will you record your findings? If it's possible, take measurements using a ruler, a tape measure or scales.

## A fair experiment

There are no right and wrong answers in science. Investigating is exciting, and during the process you may find out things you haven't thought about before.

Sometimes you'll need to make the test fair. For example, if you're testing which balls bounce the highest, you'll need to find out what different things could affect the result. These are called variables. In each fair test, you should only change one variable, and keep the others the same.

# Possible variables

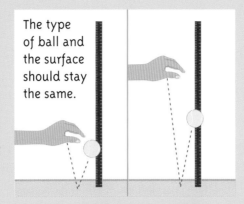

The type of ball and the surface should stay the same.

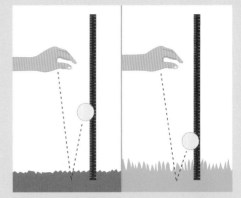

If you're testing which ball bounces the highest, the thing you'll change, the variable, is the type of ball. Everything else stays the same.

If you're testing if height makes a difference to how high a ball bounces, dropping the ball from different heights is the variable.

If you're testing the effect of different surfaces, the surface would be the variable, and the ball and height you drop it from would stay the same.

| Type of Ball | Height the ball bounced (in) |
|---|---|
| Tennis ball | |
| Soccer ball | |
| Table-tennis ball | |
| Golf ball | |

# Recording findings

There are lots of ways to record what you find out. You could take photos, make a results table, chart or graph. You could even make a poster showing what you did and what you discovered.

You could record your results in a table or a bar graph.

# Evaluation

Was your prediction correct? Write down any difficulties or problems you had. Even if an experiment or investigation didn't go as planned, it doesn't mean it was a waste of time. Learning from mistakes and redoing the experiment can often help you achieve a more accurate result. Is there anything you would change if you did your experiment again?

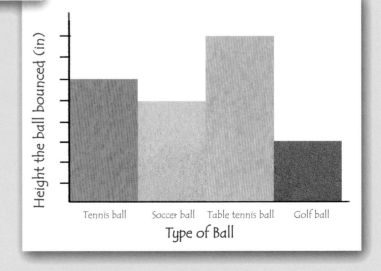

 For a link to a website where you can find out more about working on your own, go to **www.usborne-quicklinks.com**

# Glossary

This glossary explains some of the difficult or unusual words that you may have come across in this book. Any words in **bold** have their own entry in this glossary.

**airfoil** The name for a wing shape that gives an airplane an upward force.

**air pressure** The weight of air pressing down on an area.

**albumin** Chains of **chemicals** found in egg white.

**atom** One of the very tiny **particles** that everything is made of.

**barometer** An instrument used to measure **air pressure.**

**baking soda** A white powder used in cooking and mouthwashes.

**blotting paper** A type of thick paper used to absorb excess ink.

**carbon dioxide** A **gas** that is breathed out by people and animals. It is also used to make carbonated drinks.

**casein** Chains of **chemicals** found in milk.

**cell** A small unit of a living thing that contains **genes**.

**characteristic** A distinctive quality held by someone or something, such as eye color.

**chemical** A substance that is used in or results from a reaction involving changes to **atoms** and **molecules**.

**chrysalis** and **cocoon** Types of protective casings that caterpillars make. They reform inside it as a butterfly or a moth.

**compass** An instrument used for finding direction. It has a magnetized needle that points North.

**conductor** A substance through which **electricity** or heat can travel.

**cornstarch** A type of flour used in cooking for thickening sauces.

**crystal** A solid substance with a regular structure.

**density** How heavy something is in relation to its size.

**dissolve** To break up a substance into very small parts and mix them evenly in a **liquid**.

**distort** To change the appearance of something by pulling or twisting it out of shape.

**DNA** The **chemical** in all cells that is used to make up **genes**.

**droplet** A tiny drop of **liquid**.

**electric circuit** The path along which an electric current flows.

**electricity** The movement of electrically charged **particles**.

**electromagnet** A **magnet** that can be switched on and off by **electricity**.

**equator** The imaginary line that goes around the middle of the Earth.

**evaluation** An account of how well something has worked. Scientists do evaluations of their experiments.

**evaporate** To change from a liquid into a **vapor** or a **gas**.

**fiber** A long, fine thread or strand.

**force** A push or pull that makes things move, change shape or change direction.

**fossil fuel** A fuel, such as coal, oil or gas, that has formed over millions of years and is dug up from the ground.

**friction** The **force** that tends to slow down moving objects that are touching.

**gas** A substance that doesn't have a fixed shape or volume, and can fill the space it is in.

**gene** Composed of **DNA**, genes determine the **characteristics** of all living things.

**generator** A machine that turns the energy of movement into electrical energy.

**gravity** The force that keeps you on the surface of the Earth and stops you from floating away.

**horizontal** A line that goes from left to right, instead of top to bottom.

**hydroelectricity** Electricity generated from the energy of moving water.

**indicator strip/paper** A type of paper used to test whether something is an acid, an alkali or neutral. It works by changing color.

**inertia** The tendency of things to stay still, or to keep moving at the same speed, unless affected by a **force**.

**inherit** To acquire **characteristics**, such as eye color, from parents or ancestors.

**iron** A magnetic metal.

**kaleidoscope** A toy that creates light reflections through a mirrored tube.

**liquid** Something that has a certain volume, but no fixed shape. It can be poured.

**loam** A rich soil with a good mixture of clay, sand and silt.

**magnet** A material, usually made of metal, that can pull **iron** toward it.

**magnetic field** The area around a **magnet** in which its **magnetic** force works.

**magnify** To increase the size of something, as seen through a lens.

**melting point** The temperature at which a **solid** changes into a **liquid**.

**microbe** Germ or some other living thing that is too small to be seen without a **microscope**.

**microscope** Instrument used to **magnify** small objects.

**molecule** Tiny **particles** made up of **atoms**.

**nectar** The sweet watery liquid in flowers that butterflies and other insects feed on.

**nerve** Thin fiber that sends messages to and from your brain.

**nutrients** The substances absorbed by the roots of plants for nourishment.

**optical illusion** Something that your brain thinks it has seen, but which in fact does not exist.

**oxygen** A **transparent** gas in the air which humans and animals need to breathe.

**particle** A tiny piece of a material, so small that you can only see it under a powerful **microscope**.

**pinhole projector** A box with a tiny pinhole, through which you can see an image projected upside-down.

**pulse** The beat of your heart pumping blood around your body.

**rain gauge** An instrument to measure rainfall.

**reflect** The way light or sound bounces off a surface.

**reinforce** To add something to a structure to make it stronger.

**reservoir** A natural or man-made lake used for collecting or storing water.

**resting rate** The normal rate of your **pulse** before exercise.

**saliva** The **liquid** produced in your mouth that helps you to taste and swallow food.

**scatter** To spread around in various directions.

**sensors** Something that can detect changes. In the body, sensors feel things, such as heat and cold, and send signals about them to the brain.

**slack** Loose; not tight or taut.

**solid** Something that keeps its shape rather than spreading out like a **liquid** or a **gas**.

**stable** In a steady position.

**static electricity** An **electric** charge made when you rub some types of materials together.

**steel** An **iron**-based metal.

**strand** A thin thread or fiber.

**surface tension** A **force** that pulls together tiny **particles** on the surface of a **liquid**.

**swell** To get bigger.

**taste buds** A group of tiny **cells** in your tongue which enable you to sense flavors in food.

**taut** Tightly stretched.

**tornado** A swirling column of air that sometimes occurs in storms.

**transparent** Something that lets light pass through it.

**turbine** A machine that is turned by moving water or steam, and which drives **generators**.

**vapor** Another word for **gas**.

**variable** Things you change in an experiment to determine what affects what you are testing.

**vibration** A very fast backward and forward movement.

**vortex** A swirling mass or motion.

**wind vane** An instrument that shows the direction of the wind.

# List of experiments

Here you can find a list of all the experiments divided into topics.

# Index

In this index, page numbers in **bold type** show where to find the main explanation of a word or topic.

# Acknowledgements

Every effort has been made to trace the copyright holders of the material in this book. If any rights have been omitted, the publishers offer to rectify this in any subsequent editions following notification. The publishers are grateful to the following organizations and individuals for their permission to reproduce material (t = top, m = middle, b = bottom, l = left, r = right):

**p.8** © Randy Faris/CORBIS (tl); **p.20** © SPL (tl); **p.46** © Bob Rowan; Progressive Image/CORBIS (m); **p.74-75** © SCIMAT/SPL (main); **p80** © OMIKRON/SPL (tl); **p82** © BSIP, JOUBERT/SPL (tm); **p84-85** SUSUMU NISHINAGA/ SPL (main).

**Cover design**: Zoe Wray; **Art Director**: Mary Cartwright; **Digital imaging**: Mike Wheatley, Nick Wakeford and John Russell; **Additional design**: Non Figg; **Americanization**: Carrie Armstrong